AF540650

PONGAMIA
for Bioenergy and Better Environment

a division of

NIPA GENX ELECTRONIC RESOURCES & SOLUTIONS P. LTD.

New Delhi-110 034

PONGAMIA
for Bioenergy and Better Environment

M.V.R. Prasad

a division of

NIPA GENX ELECTRONIC RESOURCES & SOLUTIONS P. LTD.

New Delhi-110 034

a division of
NIPA GENX ELECTRONIC RESOURCES & SOLUTIONS P. LTD.

101,103, Vikas Surya Plaza, CU Block
L.S.C.Market, Pitam Pura, New Delhi-110 034
Ph : +91 11 27341616, 27341717, 27341718
E-mail:newindiapublishingagency@gmail.com
www: www.nipabooks.com

For customer assistance, please contact
Phone: + 91-11-27 34 17 17 Fax: + 91-11- 27 34 16 16
E-Mail: feedbacks@nipabooks.com

ISBN : 978-93-90591-48-0

Composed and Designed by NIPA.

Acknowledgement

Writing of the book on *PONGAMIA* took me a time duration much more than what is normally expected for a book of this size. This is largely due to the fact that the material on *Pongamia pinnata* was scattered and fragmentary. Most of the data and details presented in the book were worked out and generated by me during my tenure at the bioenergy company viz., M/s Vayugrid Market Places Pvt. Ltd. Bangalore, during the period from August 2009 to March 2017. In this context I must express my thanks to the staff members of the company including Mr. Doug Peterson, Mr. Gerard Rego, Mrs. Latha Chandradeep, Mr. Varun Gurjer, Mr. Chandradeep, Late Mr. Gopinath, Mr Y. Vijay Kumar, Mr. M.D. Kumar, Mr. Gopal Krishna Hegde, Mr Anil Kumar, Mr. N. Manjunatha and Mr Umashankar whose cooperation was elicited by me at different levels during the course of my work on Pongamia at Vayugrid. I am also thankful to Dr Appu Rao and Dr (Mrs) Sindhukanya of the Central Food Technological Research Institute (CFTRI) Mysore who carried out the oil analysis and biochemical investigations on seed materials of diverse elite genotypes of *Pongamia* developed by me during my tenure at Vayugrid. Also, my thanks are due to Dr. Benjamin Warr, CEO, Better World Energy, Zambia, who always interacted with me and used the elite *Pongamia* genotypes to ameliorate the toxic copper mine dumps in Zambia.

I shall be failing in my duty, if I don't express my gratitude to Dr M. Muralikrishna, who encouraged me always and steered my thinking regarding the potential of *Pongamia* on its quality parameters and processing apart from introducing Vayugrid to CFTRI and making very useful suggestions in the finalization of the draft of the book.

I am extremely grateful to Dr K. Muralidharan who took enormous interest in studying in detail my rough draft of the book and updated it with very useful information from time to time.

I thank Dr V. Ranga Rao for his useful suggestions and advice during the preparation of the text of the book. I also thank the Director and scientists of the Sardar Swaran Singh National Institute of Renewable Energy, Kapurthala (Punjab), Govt. of India, for their interest and support.

My thanks are due to M/s. New India Publishing Agency, New Delhi for publishing the book.

Last but not the least, I express my deep sense of gratitude to my esteemed Guru Prof. Dr M.S. Swamianthan for his invaluable foreword to the book on *Pongamia.*

It shall be failing on my part, if I don't acknowledge the abundant Grace and Love that I received form Shree Hari-Sai, without which it wouldn't have been possible for me to accomplish this task.

M.V. R. PRASAD

M.S. SWAMINATHAN RESEARCH FOUNDATION

M.S. Swaminathan
Founder Chairman
Ex-member of Parliament (Rajya Sabha)

Foreword

The current socio-economic situation of India presents complex problems of environmental pollution and decline of farm income coupled with rural poverty.

There are no simple solutions to the above issues; but a careful analysis of the available options does suggest some silver-linings amidst the rapidly darkening horizons. One such ameliorating options would be gradual restoration of soil fertility and enhancement of agri-ecosystems by including productive tree plantations in rural scenario.

Pongamia pinnata, popularly known in India as *Karanj* has been found to be one of the most sustainable trees with regard to its soil enhancement and carbon sequestration properties apart from its stability in seed and oil yield. *Pongamia pinnata* is also credited with several preventive and curative properties as established by Ayurvedic medicine. It may be recalled that in the decades of nineteen seventies a few villages reaped sustainable incomes in the face of acute and chronic droughts that plagued the nation, solely due to the availability of some old *Pongamia / karanj* tree stands around those zones. *Karanj* oil has exhibited promise as a source of green energy.

Nevertheless, the research on improvement of *Karanj* and its management as a productive plantation is nebulous. It is heartening that the Monograph on *Karanj* by Dr. M.V.R. Prasad fills this void. Dr. Prasad has been pursuing the work on oil bearing perennial trees of which *Pongamaia pinnata* has been studied in greater detail during the decades starting from nineteen eighties to date. The Monograph describes clearly as to how *Pongamia pinnata* could be harnessed to exploit the proven and potential benefits cited above, in addition to giving valuable information on its genetic improvement and plantation management.

This Monograph on *Pongomia* will serve as a valuable reference book for the Agricultural Scientists, Extension workers, students as well as farmers.

M. S. Swaminathan

Contents

Pongamia for Bioenergy and Better Environrment

M.V.R. Prasad

1. PONGAM TREE

Pongamia pinnata or ***Mellettia pinnata*** or ***Pongamia glabra*** or ***Derris indica*** *(*2n=2x=22*)* is a tropical ever green fast growing leguminous tree, with glabrous, glossy and deciduous leaves. The tree is native to India and South Asia and also found in Oceania. The common names of the tree in English are **Pongam** and **Indian beech**. The name Pongamia is also used as a common name without the specific name *pinnata*. In India there are several local names of the tree depending upon the region. Some of the local names are **Karanj** (Hindi), **Pungam**(Tamil), **Honge** (Kannada) **Panigrahi** (odissi) and **Gaanuga / Kaanuga** (Telugu). It is often planted as an ornamental and shade giving tree on roadside.

Pongamia is a medium-sized tree that generally attains a height of about 8.0 to 12.0 m and a trunk diameter of more than 50-75 cm. The bark is thin, grey to greyish-brown and yellow on the inside. The alternate, compound pinnate leaves of the tree consist of 5 or 7 leaflets, which are arranged in 2 or 3 pairs, and a single terminal leaflet. Pods are elliptical, 3-6 cm long and 2-3 cm wide, thick walled, and usually contain a single seed. Seeds are 10-20 cm long, flat, oblong, and light brown in colour. Seeds yield a non-edible oil and a cake is rich in protein.

Pongamia thrives in areas having an annual rainfall ranging from 400 to 2500 mm. In its natural habitat, the maximum temperature exceeds 38°C and the minimum can be as low as 1°C. The root system of Pongamia is thick and well developed deep tap-root of 10m depth and a spread of later roots to around 9m. *Pongamia pinnata* is known for its ability to fix atmospheric Nitrogen in association with soil *Rhizobia.* Mature tree apart from its inherent drought resistance, can withstand water logging and slight frost. This species grows at elevations of 1200 m, but in the Himalayan foothills it is not found above 600 m. It can grow on most soil types ranging from stony to sandy to clayey, including dry sands and saline soils. The natural distribution of this species is along coasts and river banks in India and Myanmar.

2. DISTRIBUTION

Pongamia pinnata has a broad distribution across Asia including the Indian subcontinent, Africa, Pacific and America including the Caribbean zone. Dense tree populations of *Pongamia* occur in India, Bangladesh, Pakistan, Nepal, Sri Lanka, Myanmar, Thailand, Vietnam, Malaysia, Brunei, Java, Sumatra, Singapore, Philippines, China (Fujian, Guangdong), Taiwan and Japan (Kyushu, Ryukyu Islands)).

Within India, Pongamia occurs in the states viz., Andhra Pradesh, Arunachal Pradesh, Assam, Bihar, Dadra and N. Haveli, Delhi, Goa, Gujarat, Haryana, Himachal Pradesh, Jammu and Kashmir, Karnataka, Kerala, Madhya Pradesh, Maharashtra, Manipur, Meghalaya, Mizoram, Nagaland, Orissa, Pondicherry, Punjab, Rajasthan, Sikkim, Tamil Nadu, Telangana, Tripura, Uttar Pradesh and West Bengal. It is also found in the Andaman and Nicobar Islands of Indian Union, northern Australia (including Christmas Island), Bismarck Archipelago, Papua New Guinea, Mauritius, Reunion, Seychelles, Fiji, Tonga, Samoa, Northern Marianas, Solomon Islands, New Zealand, Sudan, Egypt, Djibouti, Tanzania, Zaire, Uganda, New Zealand, Caribbean, Nicaragua and the United States of America. Across this range, its status as either native or introduced and naturalised is unclear due to a long history of local adaptation and transportation spanning back to at least the 19th Century. For example, it is considered naturalised in the United States (Florida), Puerto Rico, Hawaii and Africa (USDA, NRCS 1999). *Pongamia* was introduced to Hawaii in the 1860s and the US Department of Agriculture received seeds from Sri Lanka, Mauritius, Egypt and India in the first two decades of the 20th Century (Morton 1990).

There is uncertainty regarding the native/naturalised status of *Pongamia pinnata* in Australia. A number of publications consider *Pongamia* to be naturalised (non-native) in Australia as suggested by the Agroforestry Tree Database (Orwa *et al.*, 2009), CAB International (2000) and the Centre for Integrative Legume Research (undated). Other literature considers it native (Hnatiuk 1990; Morton 1990; Low and Booth 2007). The Queensland Herbarium currently lists *Pongamia pinnata* as native to Queensland (from Mackay north) (T Bean-Queensland Herbarium, pers. comm. 2010), with a considerable number of specimens being collected over coastal eastern and northern Queensland and the Top End of the Northern Territory. There are relatively few records south of Mackay and it is suggested that the species was planted in recent times as an ornamental south of Mackay. It is widely accepted that the tree has its origin in India and South Asia. Of-late the name *Millettia pinnata* is also being used widely in the place of *Pongamia pinnata*. There are several names given to *Pongamia*; but those are very sparsely used.

3. HABITAT, CLIMATE AND ADAPTATION

Pongamia prefers humid tropical and subtropical climates. However, it can tolerate a range of different conditions with mean annual rainfall between 400-2500 mm and temperatures of 0–16 °C minimum and 27-50 °C maximum. Mature trees can cope with light frosts, but require a dry period of 2-6 months (Duke 1983; Daniel 1997; Orwa *et al.,* 2009).

Pongamia can tolerate a wide range of soil types including saline, alkaline, sandy, heavy clay and rocky soils and waterlogged soils. However, Orwa *et al.,* (2009) suggested that it performs best in deep, well-drained, sandy loams with adequate moisture. It does not grow well in very dry sands.

Pongamia can grow at altitudes from sea level to approximately 1200 m (Orwa *et al.,* 2009). It has been described as a 'maritime species' since it tends to occur naturally along coasts and riverbanks in India, Bangladesh and Myanmar (Daniel, 1997; CAB International 2000). In Australia, it shows a preference for coastal areas and waterways (Queensland Herbarium, pers. comm. 2010).

Pongam is drought resistant tree; but can be grown under wide ranging climatic conditions too. Under scanty rainfall conditions of even less than 400mm Pongamia tree survives well and gives reasonable kernel yield levels; but thrives very well under conditions of copious rainfall up to even 2500 mm.

Fig. 1. Normal Pongamia tree population producing pods after 6 years of planting on a barren gravelly soil unfit for crop production in Mahaboobnagar Zone (Telangana) of India.

Pongamia tolerates all kinds of soil aberrations including high soil salinity, moderate alkalinity, waterlogging and poor soil fertility (Orwa et al 2009). The tree can be grown on marginal soils and conditions of low rainfall (Duke 1981 and 1983). Nevertheless, it is advisable to get the soil analysed before planting *Pongamia* in order to have a package of practices to enhance its kernel yield. The tree can't tolerate atmospheric temperatures less than 5°C and more than 60°C.

Pongam is eco-friendly and non-invasive plant species (Terviva 2006). It is known to fix atmospheric Nitrogen through its root system in symbiotic association with soil Rhizobia and enhances native soil fertility and organic matter. The tree can produce root system deeper than 10m and wider than 9 m.

Published reports on the acreage and production of *Pongamia pinnta* in India are too scanty to mention. It is estimated that the annual production of *Pongamia* kernels in India is of the order of 130,000 tons (Rabab *et al*., 1997).

An effort was made to gather the area under Pongamia in India through discussions with various organizations pertaining to Forestry, Agro-forestry and Dryland Agriculture. The discussions have revealed that the area under *Pongamia* in India is of the order of 500,000 ha in the form of unorganized tree stands, with an estimated domesticated tree population of 37.5 million (Prasad unpublished data) all along the country's rural belt. Assuming an average minimum yield of 10 kg of pods or 4 kg of kernels per tree, the total estimated pod production should be of the order of 370,000 tons resulting in an approximate total kernel / seed production of 148,000 to 150,000 tons per annum. It may be noted that the trees are not growing under any system of tree management; but are totally neglected without any care. (Anonymous, 2012).

4. TAXONOMY

Kingdom : Plantae

Phylum : Tracheophyta

Class : Magnoliopsida

Order : Fabales

Family : Fabaceae

Genus : *Pongamia*

Species : *Pongamia pinnata* Pierre

Synonyms : *Cytisus pinnatus L.* = *Dalbergia arborea* Wild = *Derris indica* (Lam) Benn. = *Derris indica* var. *xerocarpa* (Prain) Bennet = *Galedupa indica* Lam. = *Galedupa pinnata* (L) Taub. = *Galedupa pungum* J.G. Gmel = *Milletia pinnata* (L) Panigrahi,

1989 = *Pongamia glabra* Vent. = *Pongamia glabra* var. *minor* Benth. = *Pongamia glabra* var. *xerocarpa* (Hassk.) Prain = *Pongamia mitis* (L.) Kurz. = *Pongamia mitis* var. *xerocarpa* (Hassk.) Merr. = *Pongamia pinnata* var. *minor* (Benth.) Domin. = *Pongamia pinnata* var. *xerocarpa* (Hassk) Alston = *Pongamia xerocarpa* Hassk. = *Petrocarpus flavus* Lour. = *Robinia mitis*

5. POTENTIAL USES OF *PONGAMIA*

Pongamia is valued for its non-edible seed oil. The tree also yields wood used for making agricultural implements in the rural areas. The tender stems, leaves and seed cake are used as organic manures particularly in rice and sugarcane fields. The oil with leaf and seed extracts are also used to control insect pests in crops. *P.pinnata* is one of the many plantswith diverse medicinal properties where all its parts have been used in traditional medicine for the treatment and prevention of several kinds of ailments such as piles, skin diseases and wounds. (http://doi.org/10.1016/j.jep.2013.08.041 Get rights and contents).

According to the National Innovation Foundation of India, ten patents have been registered on its medicinal applications mainly for hair care and skin diseases. '*Erina Plus gel*' a product developed from *Pongamia pinnata* acts as a stimulant and helps in increasing the blood supply to skin. It prevents hair loss and skin disorders. (http://nif.org.in/PONGAMIA-PINNATA-L) The seed oil is used as *stomachic* (a medicine that serves to tone the stomach) and *cholagogue* (medicinal agent which promotes the discharge of bile from the system, purging it downward) in the treatment of dyspepsia and cases of sluggish liver. (http://www.worldagroforestry.org/)

It is used externally as liniment for treating skin diseases and rheumatic joints. It has been shown to be effective in enhancing the pigmentation and health of skin affected by leukoderma or scabies or psoriasis.

Pongam is a preferred plant species for controlling soil erosion and binding sand dunes since it produces extensive network of lateral roots and fixes atmospheric nitrogen. As it tolerates soil salinity, it is an ideal tree species for reclaiming a variety of waste lands. It is also recommended for reforestation of marginal and waste lands. Pongamia being an inclusive plant species, it promotes growth of companion vegetation such as grass beneath, it is planted to provide shade in pastures. (http://tropical.theferns.info/viewtropical.php?id=Pongamia+pinnata)

Pongamia pinnata is an important nitrogen-fixing evergreen tree with multifarious uses. Sangwan *et al.,* (2010) have reported wide ranging uses of *Pongamia pinnata* in pharmacology. It was suggested that the tree possesses anti-inflammatory, anti-diarrhoeal, anti-oxidant & anti-hyper ammonemic, anti-ulcer, anti-hyper glycemic and anti-lipid peroxidative properties in addition to the fact that the oil of the tree forms a good feed-stock for bio-energy. According to Yadav *et al.,* (2011) the roots of *Pongamia pinnata* are good for cleaning foul ulcers, cleaning teeth, strengthening gums and gonorrhoea. The root paste is used for local application in scrofulous enlargement. The fresh bark of *Pongamia pinnata* is sweet and mucilaginous to taste, soon become bitter and acrid. It is antihelmintic and useful in beri-beri, ophthalmology, dermatopathy, vaginopathy, and ulcers. Pongamia leaf, oil and seed extracts form good tools in crop protection programmes due to presence of insect toxic product karanjin (Kumar *et al* 2006).

Leaves of *Pongamia pinnata* are digestive, laxative, antihelmintic and are good for diarrhoea, leprosy, dyspepsia and cough. Flowers are useful to quench dipsia in diabetes and for alleviating *vata* and *kapha*. The seeds are antihelmintic, bitter, acrid, haematinic and carminative. They are useful in curing inflammation, chronic fevers, anaemia and hemorrhoids. The oil is antihelmintic, styptic and recommended for opthalmia, leprosy, ulcers, herpes and lumbago. Its oil is a source of biodies.

It is indigenous to the Indian subcontinent and South-East Asia and it has recently been recognised as a viable feedstock of oil for the growing biofuel industry (Karmee and Chadha 2005). The source of its oil at present is naturally growing and non-descript tree complex and very few young plantations (Divakara and Das 2011). The leaves and twigs of Pongamia are incorporated as green manure in rice fields of South India (www.soilmanagementindia.com/nutrient-elements-in-soil/green-manure-crops-techniques-and-advantages/3741).

In Florida, TerViva (2020) is planting patented varieties of the *Pongamia* trees, where the citrus industry has dramatically shrunk and few alternative crops have emerged. The company said, while in Hawaii, the trees are being planted on former sugar cane lands on Oahu, Kauai, and Maui. TerViva has 150,000 trees under contract with existing customers that include several of the largest citrus farmers in Florida and will use funding to deliver an additional 200,000 trees in the coming two years.

TerViva (2020) explained that the *Pongamia* tree products have been harvested for mostly medicinal uses for more than 1,000 years. As a plant protein, pongamia has excellent potential as a replacement for soy. *Pongamia* protein has strong gelling and emulsification properties and *Pongamia* vegetable oil is similar to high-oleic acid vegetable oils.

TerViva said it has conducted studies utilizing pongamia protein as an animal feed ingredient with poultry and cows, with promising results emerging as it prepares for regulatory submissions with the U.S. Food & Drug Administration. As a new feed ingredient, the Association of American Feed Control Officials (AAFCO) would also need to approve an ingredient definition, which would be published in the AAFCO *Official Publication*.

TerViva board chairman Ron Edwards said, "TerViva will invest this Series D fundraising support alongside Florida's citrus growers eager to revitalize fallow acreage and scale-up a sustainable, transparent oilseed supply chain. This will unlock numerous applications of the company's intellectual property growing high-performing trees and processing beans in to nutritious food and feed products to meet a broad range of needs."

Formed in 2010, TerViva is a Series-D agricultural technology company that produces *Pongamia* trees with abundant oil and protein-rich beans, similar to soybeans. It provides patented high-yielding trees and offer proprietary bean processing to create sustainable food and fuel. TerViva uses *Pongamia* to restore farmland to productive use, helping farmers feed people while taking care of the planet.

6. DESIRABLE ATTRIBUTES OF PONGM TREE

Normal Pongamia becomes a large tree within ten years with a 10-meter taproot, with substantial carbon sink. Following are some of the desirable attributes of the tree:

- Can fix more carbon than is used in production of fuel - creating a truly "carbon negative" solution.
- Resistant to drought (Fig.1), light frost, water logging, moisture stress and salinity.
- Relatively resistant to pests and diseases without much need for the use of plant protection chemicals
- Tolerant of poor soils and does not compete for prime arable land otherwise used for food production (Fig.1)
- Capable of sourcing water and nutrients well down in the subsoil.
- It is Nitrogen Fixing Tree.
- Pongamia oil yield increases until the 20th year.
- Has a lifespan of 100 years with a known productive lifespan of 60 years.
- Can be managed by a smaller/unskilled workforce due to lower crop maintenance and ability.
- A legume which needs least irrigation and minimum levels of applied nutrition.

- Regularly produces 400 – 600 kgs of seed kernels per tree per year after reaching 8 to 10 years.
- Thrives in temperatures from 0°C – 50°C

Normal unselected Pongamia tree yields around 4 to 7 kg/tree after the 5th or 6th year of age. A genetically elite graft of Pongamia offers a production of an average of 80kg / tree / year from year-10, throughout its life of 100 years.

- Ten trees can yield 400 litres of oil, 1200 kg of fertilizer grade oil cake and 2500kg of biomass as green manure per year. Manurial value of leaves and twigs of *Pongamia* are given Table.1.

An evaluation of the wide range of plants such as oil yielding plant species for bio-diesel production has indicated that *Pongamia pinnata* (aka *Mellittia pinnata)* has all the desirable attributes that render it suitable to be grown with profit under the above mentioned eco-systems. A comparison of *Pongamia* (*Mellittia pinnata)* with another plant species viz., *Jatropha curcus* a candidate for bio-diesel production is given below. Oil Palm also figures in the Table for comparison (Table 2 and Fig.2)

Table 1. Manurial Value of *Pongamia.*

Nutrients	LEAF	TWIG
N (%)	1.16	0.71
P205 (%)	0.14	0.11
K20 (%)	0.49	0.62
CaO (%)	1.54	1.58

(*Source:* http://www.millettiaplantations.com/botanical-overview.htm)

Table 2. Comparison of Oil yielding Plantation Crops

Factor	Oil Palm	Jatropha	Millettia (Pongamia)
Rainfall requirement per year	2,000-3,000mm	600-2,000	250-2,500mm
Harvest Method	Manual	Manual	Easily Mechanized
Plantation Carbon Credits Qualified	No	No	Yes
Area managed per plantation worker	10Ha	5Ha	60Ha
Oil Yield / ha -3 years in tonnes	2.75	3	10
Oil Yield / ha - 10 years in tonnes	10.5	9.5	18
Oil Yield / ha - 15 years in tonnes	11	10	28

Yields above are based on results from plantations that are professionally managed with high quality plant genetics, adequate soil nutrition, proper harvesting techniques and high quality processing equipment.

Oil Yield - Millettia pinnata vs other oil seed crops
Tonnes/Per Hectare/Per Year @ 10 Year

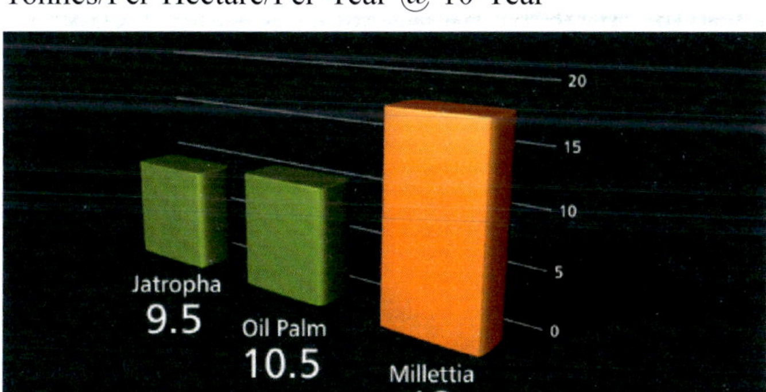

Fig. 2. Comparision of *Melletia (Pongamia) pinnata* with other oil yielder trees (*Source:* http://www.millettiaplantations.com/botanical-overview.htm)

To sum up, a higher recovery and better quality of oil of *Pongamia pinnata* as compared to other crops, its non-competition with food crops as it is a non-edible source of fuel, and its suitability to be grown on degraded and marginal lands render it a highly desirable not only in ameliorating the farmers' economy, but also enhancing the eco-system and environment (Prasad 2012a).

Studies carried out by Al Muqarrabun *et al.,* (2013) indicated that several different classes of flavonoid derivatives, such as flavones and chalcones, and several types of compounds including terpenes, steroid, and fatty acids have been isolated from all parts of this plant. The pharmacological studies revealed that various types of preparations, extracts, and single compounds of this species exhibited a broad spectrum of biological activities such as antioxidant, antimicrobial, anti-inflammatory, and anti-diabetic activities. The results of several toxicity studies indicated (Al Muqarrabun *et al.,* 2013) that extracts and single compounds isolated from this species did not show any significant toxicity and did not cause abnormality on some rats' organs.

In traditional system of Ayurvedic medicine *Pongamia pinnata* has been widely used as curative agents for variety of ailments. According to Yadav *et al.,* (2019), concentrated extract of fruits or seeds can be found in various herbal preparations widely available in market today. *Pongamia pinnata* oil preparation is widely available and employed by practioners of natural health for treatment of rheumatism. In the traditional systems of medicines, such as Ayurveda and Unani, the *Pongamia pinnata* plant is used as a source for anti- inflammatory, anti-plasmodial, anti-nociceptive, anti- hyperglycaemic, anti-lipidperoxidative, anti-diarrhoeal, anti-ulcer, anti-hyperammonic, anti-oxidant and antibacterial treatments.

Thus, this plant has a potential to be used as an effective therapeutic remedy due to its low toxicity towards mammalian cells.

Pongamia has been reported to be efficient in remediating problem soils containing heavy metals and other pollutants that impede plant growth (Prasad 2007, Tulod *et al.*, 2012; Kumar *et al.*, 2017; Prasad, 2019). The phytoremediation potential of *Karanj* has been examined in few comparative studies; in pot experiments, on coal mine spoil, fly-ash amended soils and on landfill wastes (Pandey, 2017).

The grafted clones of elite Pongamia have exhibited impressive potential with regard to oil yield (Prasad and Vijaykumar 2016). As a legume, it fixes nitrogen from the soil, minimizing the need for added fertilizers. Emergence of the new *Pongamia* based cropping system may change the entire scenario of the biodiesel industry and shall provide much relief to the industry which is desperately in a need of a viable sustainable non-food feed stocks.

Fig. 3. Young Elite Pongamia Plantation in Australia

7. GROWTH AND DEVELOPMENT

It is a fast growing, medium sized evergreen tree attaining a height of 7-10 m and stem diameter of 50-80 cm many times with drooping canopy. The tree at times can grow to a height of even 20-25 m with a trunk diameter of 60cm. It has smooth grey-brown bark with vertical fissuring Leaves compound, pinnate and alternate. Pongamia tree has the typical feature of producing sprouts from the base of the trunk and more particularly root-suckers. Mature leaves are glossy and dark green above and pale below. New leaves are pinkish-red. Flowers are white, pink or lavender pea-like blossom. Bloom occurs in late spring/early summer. Seeds are 1.5 cm long, light brown, oval and contained in clusters of brown, eye-shaped pods. The tree reproduces from viable seeds but asexual root suckers too grow into trees. Yield ranges from 9 to 90 kg seed/tree. The seeds possess oil content ranging from 25 to 40%. A major component of the seed oil is Oleic Acid, which ranges from 45 to 60% of seed oil.

8. FLORAL BIOLOGY, PODS AND WOOD

i) Floral Biology

Fig. 4. Pongamia leaves

Flowers are fragrant, white to pinkish, paired along rachis in axillary, pendent, long racemes or panicles. Calyx is cup-shaped. Tree flowers in 4th to 5th year. The flowers are paplinaceous with one standard, two wings and two keels. Stamens are free at the base, but joined with others into a closed tube. Ovary is short-stalked, pubescent with generally two ovules but rarely three. Style is filiform with upper half incurved and glabrous. Stigma is small, terminal, knob shaped white and papillate that remains receptive during 06.00 to 11.30 AM following flower opening. Completion of anthesis is indicated by the closure of the flower. *Pongamia* is dry season bloomer and produces flowers with copious amount of pollen and additionally secrets nectar which form droplets on wing and keel petals that attract bees (Raju and Rao 2006). Pollination in Pongamia is effected by several species of bees of which *Apis dorsata, A. millifera, A. cernaindica, Amegilla* spp., *Megachile* and *Xylocorpa* spp., are important.

There is a substantial out-crossing in *Pongamia* due to insect pollination. Pollen viability is of the order of 83%. After 72 hours of pollination, pollen tubes are found to traverse through stylar tissue. Visit of the honey bee viz. *Apis mellifera* is reported to increase seed yields (Arpiwi *et al*., 2014). Setting up of bee hives in Pongamia plantation for honey production could be a practical proposition (Daisy *et al.*, 2001).

Fig. 5. Blossming Flowers of *Pongamia pinnata*

ii) Pods and Seeds

Pods are short stalked, oblique to oblong, flat, smooth, thickly leathery to sub-woody and indehiscent. Pods are normally 1-seeded; but occasionally two seeded. Seeds are thick, reniform and viable for approximately one year and contain 28 to 40% of oil. Seeds germinate in 10 to 16 days after sowing. Pod shells of Pongamia form a good material for making briquettes,which could be an ideal kitchen fuel (VAYUGRID 2014)

Fig. 6. Pods & Seeds of Pongamia pinnata

iii) Wood

Wood is beautifully grained and medium to coarse textured, but is not durable. Therefore, it is used as fuel-wood and cheap timber. The leaves are not readily eaten by animals, but it has some fodder value in dry areas. Leaves are also used as insect repellent in stored grains. The oilcake has use as poultry feed, as manure with insecticidal and nematicidal value. The cake can also be used to produce biogas.

9. SEED OIL AND ITS FUEL CHARACTERISTICS

Pongamia oil is extracted from the seeds by expeller pressing, cold pressing, or solvent extraction (Berk, 2013). The oil is yellowish-orange to brown in colour. Pongamia oil is known for its insect repellent properties (Pavela and Herda 2007). However, the oil can be used to produce bio-diesel through the process of trans-esterification. The average seed oil content ranges from 25-40%, but higher percentages are claimed in elite genotypes. The yield of methyl esters from the oil under the optimal conditions was 97–98%. The oil is thick and yellow-orange to brown in colour. Traditionally, besides the cooking and lighting uses in rural areas, it was used as a lubricant, water-paint binder, pesticide, and in soap making and tanning industries. The oil is known to have value in folk medicine for the treatment of rheumatism as well as human and animal skin diseases. Seed oil is also used in scabies, leprosy, piles, ulcers, chronic fever, lever pain and lumbago. (Yadav *et al.*, 2011).

The Pongamia oil is reported to exhibit following fatty acid profile (Anonymous 2012A) (Table 3).

Table 3. Fatty acid profile of Pongamia oil

Fatty Acid	Content (%)
Palmitic acid	3.7 to 7.9
Stearic acid	2.4 -8.9%
Oleic acid	44.5 – 71.3%
Linoleic acid:	10.8 – 18.3%
Linolenic acid	2.60%
Arachidic acid	2.2 – 4.7%
Eicisenoic acid	9.5 – 12.4%
Behenic acid	4.2 – 5.3%
Lignoceric acid	1.1 – 3.5%

Straight vegetable oil too can be used as source of energy, particularly to run the village pump-sets, tractors and old heavy vehicles. The cake left out after extraction of oil forms valuable organic manure. The seed also contains protein of good quality to the tune of more than 17%. The seed cake can be used to produce bio-gas.

Pongam oil is used in tanning industry for dressing of EI leathers. In leather industry it is used for making fat liquors. After neutralization, the oil is used as lubricant for heavy lathes, chains, bearings of small gas engines, enclosed gears and heavy engines.

A study by RANWA (Research and action in Natural Wealth Administration) says the Pune Municipal Transport could run its 800 bus fleet on karanja oil if the tree is planted on 68.6 sq. km. (Anonymous 2002) The karanj oil has been used in 7.5 KVA gensets has electrified three villages in Adilabad district of Andhra Pradesh and in six villages in Jadol area of Rajasthan (Anonymous 2002). The utility of the seed oil of *Pongamia* has been well established (Shrinivasa 2001 and Vivek & Gupta 2004). De and Bhattacharya (1999) studied the fuel characteristics and found karanja oil can be used as cheap raw material for synthesis of biodiesel (Table 4). Karikalan and Chandrasekaran, (2015) studied the physical and chemical characteristics of vegetable oils and found karanja oil can be used as an efficient and cheap raw material for synthesis of Biodiesel (Table 5).

Table 4. Fuel characteristics of karanja oil (De and Bhattacharya 1999)

Parameters	Pongamia oil (unfiltered)	Diesel
Saponifcation value	178.47	Nil
Iodine value	115.8	38.3
Acid value	2.29	0.06
Colour in 1 inch cell (Y+5R)	50	102.5
Specific gravity	0.92	0.84
Refractive Index	1.481	1.472
Moisture & Volatile matter	0.05%	24.66%
Viscocity at 25°C	84.9 CST	8.5CST
Cloud point	15°C	13°C
Pour point	-1°C	1°C

Table 5. Physical and Chemical Properties of karanj oil (Karikalan and Chandrasekaran, 2015)

Properties	Unit	Test Value
Densitty	gm/cc	0.927
Kinematic viscosity @40*C	Mm square/s	40.2
Acid value	mg KOH/gm	5.4
Pour point	°C	6
Cloud point	°C	3.5
Flash point	°C	225
Calorific Value	MJ/Kg	8742
Saponification value		184
Carbon residue	Wt%	1.51
Specific gravity		0.936

According to Ahmad *et al.*, (2009) the fuel properties of biodiesel (100%) are inclusive of 0.92, kinametric viscocity of 7.53 cSt @40°C, flash point at 90°C, sulphur contents of 0.0084 (% weight), pour point of -6°C, cloud point at 4°C, distillation (initial boiling point) of 215 and cetane number of 53, which were near high-speed diesel

Fig. 7. *Pongamia* Oil and Karanjin Powder

Production of biodiesel from Pongamia oil

Biodiesel from *Pongamia pinnata* is obtained by the transesterification process. The trans-esterification process is the reaction of triglyceride (fat/oil) with an alcohol in the presence of acidic, alkaline or lipase as a catalyst to form mono alkyl ester that is biodiesel and glycerol. However, the presence of strong acid or base accelerates the conversion.

It is reported that alkaline catalysed trans-esterification is the fastest and requires only simple set up. Oil of *Pongamia pinnata* may be trans-esterified with methyl alcohol in presence of strong alkaline catalyst like sodium hydroxide (NaOH) or potassium hydroxide (KOH) in a batch type trans-esterification reaction. After trans-esterification of crude oil of pongamia, the resulting product shows excellent properties like calorific value, iodine number, cetane number and acid value etc. (Bodade and Khyade, 2012).

A maximum conversion of 92% (oil to ester) could be achieved using 1:10 molar ratio of oil to methanol at 60°C. Tetra-hydro-furan (THF), when used as a co-solvent, the conversion could be increased to 95%. Solid acid catalysts viz., H^2 -Zeolite, Montmorillonite K-10 and Zinc oxide could also be used for transesterification.

Important fuel properties of methyl esters of *Pongamia pinnata* oil (resulting in biodiesel viscosity of 4.8 cSt @40°C and flash point at 150°C) compare well with ASTM and German biodiesel standards (Kumar and Chadha, 2005). Biodiesel yield of 99% was obtained from Pongamia oil by transesterification under the molar ratio of 6:1of alcohol to oil with sodium hydroxide as a catalyst of 0.5 (vol.%) at 60°C with the completion of production process in one hour. The study showed that specific fuel consumption and thermal efficiency of B20 biodiesel from Pongamia are quite comparable to that of petroleum diesel. The hydrocarbon and carbon monoxide gas emissions are quite low in Pongamia biodiesel as compared to diesel (Dwivedi and Sharma, 2014).

10. PONGAMIA CAKE

It can't be used as a source of human or animal nutrition due to the presence of some irritants and anti-nutritional substances such as karnjin, saponins and phytates, although not being toxic. Nevertheless, it has been conclusively demonstrated by CFTRI that protein of *Pongamia* cake is comparable to that of soybean in terms of quality following removal of the above mentioned irritants through simple acid hydrolysis (Vinay and Sindhukanya, 2007; Soren *et al.*, 2009). Such conversion of *Pongamia* seed cake into animal feed with good nutritional qualities may be more useful (Panda *et al.*, 2006). It is worth attempting converting Pongamia seed cake into animal feed in view of its promising nutritional qualities.

TerViiva a USA company working on the development of Pongamia, has established a system to develop food products from Pongamia tree, thereby introducing a whole new source of plant based protein and oil source to industry (Neo, 2020).

Pongamia cake could be used as an effective organic fertilizer. It is rich in NPK: 4:0.5:0.5 and karanjin at > 800 ppm., facilitating improvement of soil fertility. The press cake when applied to soil, it has pesticidal value as it controls biotic stresses particularly nematodes. As a natural fertilizer, it can be mixed with neem cake pellets to give better result, thereby cutting down on cost of cultivation apart from ensuring crop protection from insect pests, soil borne pathogens and nematodes. Pongamia cake is being widely used on crops like rice, wheat, cotton, sugarcane, vegetables, spices, coconut, tea, coffee plantations, fruit orchards, aromatic oil yielding plants and golf grounds (Green My Life, 2017). Perceptible increase in cotton yield has been reported due to application of 100% Pongamia cake, which is higher as compared to the application of 100% of inorganic fertilizers (Osman *et al.,* 2005) (Table 6).

Table 6. Plant Nutrient Composition of Pongamia Cake (Osman *et al.*, 2005)

Plant Nutrent	**Content**
Nitrogen %	4.28
Phosphorous %	0.4
Potassium %	0.74
Calcium %	0.25
Magnesium %	0.17
Zinc (ppm)	59
Iron (ppm)	1000
Copper (ppm)	22
Manganese (ppm)	74
Boron (ppm)	19
Sulphur (ppm)	1894

Fig. 8. Pongamia cake

11. KARANJIN AND PONGAMOL

Pongamia produces natural chemical compounds, such as karanjin and pongamol, which impart bitter taste to the foliage and other parts of the tree. These compounds deter consumption of leaves by herbivores and insect infestations, resulting in lower management costs and enhanced sustainability. In addition, these compounds are valued for use in crop sprays, cosmetics, and sunscreen products. Pongamol and karanjin isolated from the pods of *Pongamia pinnata* possess significant antihyperglycemic activity in streptozotocin-induced diabetic rats and type 2 diabetic db/db mice and protein tyrosine phosphatase-1B may be the possible target for their activity (Tamarkar *et al.,* 2008)

i) **Karanjin,** a white crystalline powder is furano-flavanol (3-methoxy furano-2,3,7,8-flavone) in quantities of 2% by seed weight and 4-5% by weight of oil.

   The extraction of karanjin from seeds of Pongamia has been reported by Vismaya *et al.*, (2010)

Karanjin

Karanjin is an acaricide and insecticide. Acaricide means a product that is used to kill mites (Acarina). Karanjin is used as bio pesticide / bio insecticide and is reported to have nitrification inhibitory properties (Majumdar *et al.,* 2004). The resistance of Pongamia trees to termites and wide range of other pests is due to karajin and pongamol in the plant system. Karanjin could be used as an efficient pesticide / pest –repellent in organic agricultural programmes where chemical pest control is not allowed. (Lale and Kulkarni, 2010).

The effects of an identified antihyperglycemic molecule, karanjin, isolated from the pods of *Pongamia pinnata* were investigated on glucose uptake and GLUT4 translocation in skeletal muscle cells. Treatment of L6-GLUT4myc myotubes with karanjin caused a substantial increase in the glucose uptake and GLUT4 translocation to the cell surface, in a concentration-dependent fashion, without changing the total amount of GLUT4 protein and GLUT4 mRNA (Jaiswal etal 2011). This effect was associated with increased activity of AMP-activated protein kinase (AMPK). Cycloheximide treatment inhibited the effect of karanjin on GLUT4 translocation suggesting the requirement of de novo synthesis of protein. Karanjin-induced GLUT4 translocation was further enhanced with insulin and the effect is completely protected in the presence of wortmannin. Moreover, karanjin did not affect the phosphorylation of AKT (Ser-473) and did not alter the expression of the key molecules of insulin signaling cascade. The authors conclude that karanjin-induced increase in glucose uptake in L6 myotubes is the result of an increased translocation of GLUT4 to plasma membrane associated with activation of AMPK pathway, in a PI-3-K/AKT-independent manner. Vismaya *et al.,* (2010A) reported the gastro-prtoective properties of karanjin in addition to its anti-oxidant nature, there by suggesting its use in medicine.

ii) **Pongamol** is 1-(4-Methoxy-5-benzofuranyl)-3-phenyl-1,3-propanedione. It is a pale yellow to cream-coloured powder. Extraction of karanjin and pongamol from Pongamia oil by reverse phase HPLC has been reported by Gore and Satyamoorthy (2000). The quantity of Pongamol in Pongamia oil is much lesser than that of karanjin (Table 7).

In view of its low recovery it is stated that the cost of Pongamol is always much higher than that of Karanjin.

H_3CO O O

O

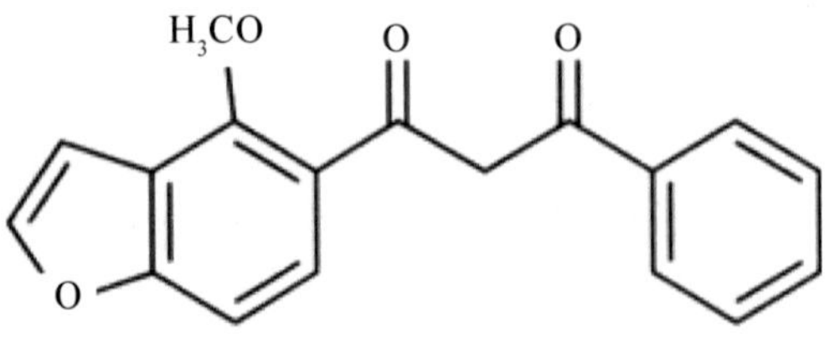

Pongamol

Table 7. Pongamol an Karanjin contents in various sources of Pongamia oil (Gore and Satyamoorthy 2000)

Oil Sample	% of Pongamol	% of Karanjin
1	0.37	2.17
2	0.49	1.12
3	0.88	4.53
4	0.9	3.67

12. CYTOLOGY AND GENETIC IMPROVEMENT INCLUDING BIOTECHNOLOGICAL INVESTIGATIONS

The somatic chromosome number of *Pongamia pinnata* is 2n=2x=22 (Patel and Narayana 1937, Raghavan and Arora 1958, Bhatt and Sanjappa 1976 and Sarbhoy 1977). Singhal *et al.,* (1990) reported existence of B chromosomes to the tune of 7 numbers in the somatic cells of *Pongamia pinnata.* The mitosis and meiosis of Pongamia is more or less regular with presence of one or two quadrivalents in meiotic Diakinesis to Metaphae-1 as shown by Sarbhoy (1977) (Fig. 9).

Besides these cytomorphological variables, structural heterozygosity for translocations/inversions are also reported. Bhatt and Sanjappa (1976) found that the species has five pairs of long chromosomes (3.06-3.90 mm) of which three pairs have centromeres which are approximately median and two pairs have centromeres which are approximately sub-median. The six pairs of short chromosomes (1.70-2.55 mu m) have sub-median centromeres. Eleven bivalents occur at meiosis, and laggards, bridges and cytomixes are common abnormalities probably due to the presence of B chromosomes. Consequently, the percentage pollen fertility was 56.40%.

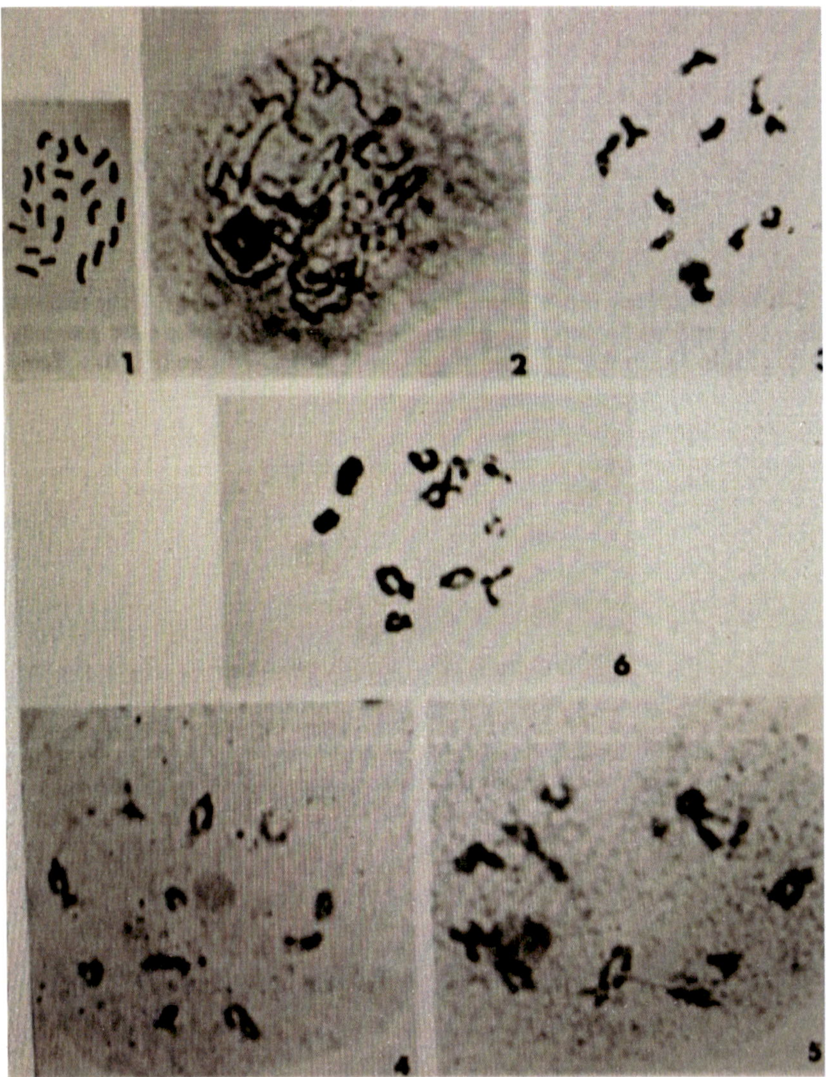

Fig. 9. Plate1. Mitosis in *Pongamia pinnata*. Metaphae showing 2n=22 chromosomes. **Plates. 2 to 6** Meiosis in *Pongamia* with various phases of Diakinesis. See one quadrivalent and 9 bivalents in late Diakinesis in **Plate No. 5.** and in **Plate No. 6**, Note 2 quadrivalents and 7 bivalents. [*Source*: Sarbhoy,1977]

Kesari *et al.*, (2008) found that candidate plus trees (CPTs) of *Pongamia pinnata*, occurring across 10 locations in North Guwahati were identified based on morphological markers (vegetative and reproductive) using combined analysis over locations. Identified CPTs were then multiplied using seed propagation technique in a nursery bed. The performance of the candidate trees with respect to seed and pod traits, the two most important characters with regard to oil, were evaluated using CROPSTAT software for inferring potential genotypes that can be included in programmes aimed at genetic improvement of the species. Total oil content from the seeds of plus trees was also analysed using solvent extraction procedure at their boiling points. Hexane extraction yielded maximum oil content from seeds (33%) compared with petroleum ether (30%). When the seed to solvent ratio varied, no significant difference was noticed on the total oil yield for an individual tree, although the recovery of solvent and the time taken for oil extraction were significantly reduced at higher ratios of solvent used.

According to Mukta *et al.*, (2009) wide variability in oil content was observed in 75 germplasm accessions of *Pongamia pinnata* (L.) collected from Telangana in India. Out of these, fatty acid profiles of 21 accessions with varying seed oil content were examined. Large variation was observed in stearic, oleic and linoleic fatty acid composition i.e. 1.83–11.50%, 46.66–65.35% and 12.02–32.58% respectively while less variation i.e. 9.25–12.87% was found with palmitic acid content. Saponification number (SN), iodine value (IV) and cetane number (CN) of fatty acid methyl esters of oils varied from 183.3 to 200.91, 74.78 to 100.98 and 50.85 to 59.11 respectively. Fatty acid composition, IV and CN were used to predict the quality of fatty acid methyl esters of oil for use as biodiesel. Fatty acid methyl esters of oils of *P. pinnata* accessions DORPP 49, 72 and 83 were found most suitable (CN more than 56.6) for use as biodiesel and they meet the major specification of biodiesel standards of USA, Germany and European Standard Organization. The range of variability found for various biodiesel standards in accessions of *P. pinnata* can be utilized for the establishment of plantations of promising genotypes through clonal means for increased productivity.

Sunil *et al.*, (2010) studying the Variability and divergence in the germplasm of *Pongamia pinnata* collected from Karnataka (India), found good genetic diversity and variability for pod, seed and other characters of bio-productivity and suggested that the selection of the candidate plus (elite) trees for genetic improvement of pongam tree would be necessary to achieve progress in seed and oil yield.

Sahoo *et al.*, (2011) reported that the candidate plus trees (CPTs) selected from different zones of Orissa state in India showed significant variation among themselves in respect to their pod and seed characters. Phenotypic coefficient of variation (PCV) and genotypic coefficient of variation (GCV) estimates were high for pod thickness, seed thickness, 100-pod weight, and 100-seed weight. High heritability values accompanied by high genetic advance for 100-seed weight (96.1%, 59.6) and 100-pod weight (90.9%, 37.3) indicated additive gene action. High estimates of genotypic correlations than the corresponding phenotypic correlations indicated the presence of strong inherent association between pod length and pod breadth; 100-pod weight, and pod thickness; 100-pod weight and seed length; 100-seed weight and 100-pod weight. Seed length, seed breadth, seed thickness, 100-pod weight and 100-seed weight had significant positive correlation with each other, and these characters should be considered as effective parameters to select CPTs for different agroforestry programs.

In the recent years, *Pongamia* oil industry has caught the great attention in the global market and, since India, one of the world's Pongamia producers plays a very important role in it. Thus, it is very important for stakeholders of *Pongamia* oil industry to continuously and constantly to keep updated on the technology due to its potential in maximizing productivity and achieving further efficiency.

Centre for Jatropha Promotion (www.jatrophabiodiesel.org) has reported to have got an improved highly yielding and early maturing variety of Pongamia and developed enhanced crop cultivation technology with oil yield of 330 gallon in 4th year, 1000 gallon in 7th year and 2000 gallon in 10th year from 1 ha plantation.

Prasad (2013) reported that elite trees of Pongamia selected based on a set of stable attributes associated with yield from out of the natural variable tree populations in southern Karnataka (India) exhibited consistently higher degrees of seed and oil yield. The grafted clones of the reported elite trees indicated immense promise for enhanced recovery of kernel and oil yields (Prasad 2012b and 2013).

Impressive genetic diversity of *Pongamia* with wide array of variability for wide range of characters in India and South-east Asia provides ample scope for genetic improvement of the species for the economic attributes viz., seed yield and oil content. Without genetic enhancement of *Pongamia*, it would not be possible to develop promising planting material for feed stock for bio-diesel production.

Biotechnological Investigations: The core of biotechnology is the ability to identify, isolate, amplify and modify genes or sequences of DNA from any source and to deliver and express the novel modified sequences in the same or different species.

It includes the development of specific organisms for specific application or purposes and may include the use of novel technologies such as tissue culture, recombinant DNA, cell fusion, and other new bioprocesses.

Improvement of woody tree species by traditional plant breeding techniques has several limitations mainly caused by their high degree of heterozygosity, the length of their juvenile phase and auto-incompatibility. Vegetal improvement through biotechnological approaches is mostly concerned with tissue culture for regeneration of novel genetic material, protoplast fusion to get somatic hybrids, gene transfer to get genetically modified organisms and use of DNA markers to select trait of interests. Varieties with improved biotic and abiotic stress resistance is expected to be developed in less time and with more accuracy using recent biotechnological approaches. Several advance tools could be utilized for the purpose including, nanotechnology, bioinformatics tools etc., that promise a new era of genomics assisted molecular breeding. Next Generation Sequencing and high throughput genotyping approaches may increase efficiency and output of biotechnological tools in agriculture. The development of new biotechnological tools (NBTs), such as RNA interference (RNAi), trans-grafting, cisgenesis/ intragenesis, and genome editing tools, like zinc-finger and CRISPR/Cas9, has introduced the possibility of more precise and faster genetic modifications of plants. This aspect is of particular importance for the introduction or modification of specific traits in woody tree species while maintaining unchanged general characteristics of a selected cultivar. Moreover, some of these new tools give the possibility to obtain transgene-free modified tree genomes, which should increase consumer's acceptance. Over the decades, biotechnological tools have undergone rapid development and there is a continuous addition of new and valuable techniques for plant breeders. This makes it possible to create desirable woody tree varieties in a fast and more efficient way to meet the demand for sustainable agricultural productivity. Although, NBTs have a common goal i.e., precise, fast, and efficient crop improvement, individually they are markedly different in approach and characteristics from each other.

The biotechnology of *Pongamia* is still in its infancy due to the general neglect of this important species of tree and lack of basic genetic data. Some investigations have been carried out in the areas of tissue culture and regeneration technology in India and Australia. Despite the fact that the tissue culture work in *Pongamia* dates back to 1997, most of the earlier reports were on callus induction (Rajan 1997). The later investigations focused on tissue culture work and the pathways of morphogenesis in *Pongamia*. Notwithstanding the slow response of the plant material and mature explants requiring elaborate sterilization procedures, some success has been reported for whole plant regeneration and hardening through de novo caulogenesis (Sujatha et al., 2008) and through micro-propagation (Sujatha and Hazra, 2007).

Sugla et al. (2007) working on micro-propagation of *Pongamia pinnata*, induced multiple shoots in vitro from nodal segments through forced axillary branching (Fig.10). According to them, Murashige and Skoog (MS) medium supplemented with 7.5 ¼ M benzyl-aminopurine (BAP) induced up to 6.8 shoots per node with an average shoot length of 0.67 cm in 12 d. Incorporation of 2.5 ¼ M gibberellic acid (GA3) in the medium during the first subculture after establishment and initiation of shoot buds significantly improved the shoot elongation.

Fig. 10. Multiple shoot production from a cotyledonary node of *Pongamia pinnata.* (Sugla *et al.*, 2007)

Single use of GA3 during the first subculture eliminated the need for prolonged culturing on BAP medium. Further use of GA3 in the medium, however, was not useful.

Shoot culture was established for at least two subcultures without loss of vigor by repeatedly sub-culturing the original cotyledonary node on shoot multiplication medium followed by shoot elongation medium after each harvest of the newly formed shoots. Thus, from a single cotyledonary node, about 16-18 shoots were obtained in 60 d.

Shoots formed in vitro were rooted on full-strength MS medium supplemented with 1.0 ¼ M Indole butyric acid (IBA). Plantlets were successfully acclimatised, established in soil, and transferred to the nursery.

Sujatha et al. (2008) achieved the regeneration of *P. pinnata* via adventitious organogenesis using de-embryonated cotyledons as explants with thidiazuron (TDZ) at different concentrations.

Kesari et al. (2010) studied regeneration efficiency and genetic-clonality for in vitro propagation of candidate plus tree of Pongamia pinnata. Woody Plant Medium (WP) and Murashige and Skoog's medium (MS) supplemented with different concentrations and combinations of plant growth regulators were screened for high frequency regeneration using nodal segment culture and axenically grown seedlings of an elite genotype of *P. pinnata*. Percentage response from field-grown mature nodal segments of *P. pinnata* were highly dependent on the season, with greater than 68% of culture developing adventitious shoots during spring. Woody Plant Medium supplemented with benzyl-adenine (5.0 mg L^{-1}) and kinetin (0.5 mg L^{-1}) gave the greatest response to initiation and multiplication. The multiplication rate of 11 shoots per explants with an average shoot length of 3.0 cm was observed. Multiplied shoots started to produce roots in the multiplication medium itself containing BA and NAA; but subsequent establishment was poor. The rooting response was enhanced in half-strength MS media with indole-3-butyric acid (0.5 mg L^{-1}). Rooted plants were hardened successfully in glass house with 70% survivability. RAPD and ISSR markers were employed to determine the genetic fidelity of *in vitro* raised plantlets.

Fig.11. Regenerated sapling of *Pongamia* from callus of cotyledon (Biswas *et al.*, 2013)

According to Biwas et al., (2013), regeneration by organogenesis from nearly mature green cotyledons has been demonstrated using the plant hormone thidiazuron [1-phenyl-3-(1,2,3-thiadiazol-5-yl) urea] and a regeneration frequency of 66% has been reported. However, success of this procedure is highly dependent on genetic background of each individual tree as demonstrated by differing responses of explants from different candidate trees (Brisbane street trees) under identical tissue culture conditions. Whereas some plants were able to be regenerated quite efficiently (up to 100%) from cotyledonary explants, others lacked even basic callus formation. Choudhury et al., (2013) carried out biotechnological research at the Indian Institute of Technology in Gauhati (India). The research brought out that nuclear haploid DNA content of 1,300 million base pairs for *P. pinnata* will be a universal reference for biofuel crops. The research lead to the discovery of key gene involved in fatty acid biosynthesis. Also new species *Rhizobium pongamiae* from root nodules of Pongamia- salt tolerant (4%), temperature resistant (max 42 degrees C) and alkaline pH has been developed.

Shelke (2010) studied genetic variation in Pongamia population in Konkan region with Polymeric Chain Reaction- Restriction Fragment Length Polymorphisms (PCR-RFLPs) and Chloroplast Simple Sequence Repeats (cpSSR). Based on the study further tree improvement program was suggested.

Jin *et al*., (2019) worked on role of the MicroRNA-mediated gene Regulation in developing Pongamia seeds by small RNA profiling. According to them, the microRNAs (miRNAs) are important molecular regulators of plant development; but relatively little is known about their roles in seed development, especially for woody plants. During the course of their investigation, 236 conserved miRNAs were identified within 49 families and 143 novel miRNAs via deep sequencing of Pongamia seeds sampled at three developmental phases. The miRNAs, 1327 target genes were computationally predicted for these cases. Additionally, 115 differentially expressed miRNAs (DEmiRs) between successive developmental phases were sorted out. The DEmiR-targeted genes were preferentially enriched in the functional categories associated with DNA damage repair and photosynthesis. The combined analyses of expression profiles for DEmiRs and functional annotations for their target genes revealed the involvements of both conserved and novel miRNA-target modules in Pongamia seed development. Quantitative Real-Time PCR validated the expression changes of 15 DEmiRs as well as the opposite expression changes of six targets. The reported results provide valuable miRNA candidates for further functional characterization and breeding practices in Pongamia and other oilseed plants.

One of investigations on the biotech applications to *Rhizobium* spp. has resulted in the identification of a gram-negative, non-motile, fast-growing, rod-shaped, bacterial strain VKLR-01T which was isolated from root nodules of Pongamia that grew optimal at 28°C, pH 7.0 in presence of 2% NaCl. Isolate VKLR-01 exhibits higher tolerance to the prevailing adverse conditions, for example, salt stress, elevated temperatures and alkalinity. Strain VKLR-01T has the major cellular fatty acid as C18:1 É7c (65.92%). Strain VKLR-01T was found to be a nitrogen fixer using the acetylene reduction assay and PCR detection of a nifH gene. On the basis of phenotypic, phylogenetic distinctiveness and molecular data (16S rRNA, recA, and atpD gene sequences, G + C content, DNA-DNA hybridization etc.), strain VKLR-01T = (MTCC 10513T = MSCL 1015T) is considered to represent a novel species of the genus *Rhizobium* for which the name *Rhizobium pongamiae* sp. nov. is proposed. *Rhizobium pongamiae* may possess specific traits that can be transferred to other rhizobia through biotechnological tools and can be directly used as inoculants for reclamation of wasteland; hence, they are very important from both economic and environmental prospects (Kesari et al., 2013A).

Genetic engineering technologies in plant improvement have been in practice for more than three decades. Direct transformation methods (Biolistic) and indirect methods (*Agrobacterium tumefaciens*-mediated transformation), developed decades ago, have been the primary strategies of heterologous DNA introduction into plants (Gelvin, 2003; Altpeter *et al.*, 2005). All genetically modified crops commercially grown, including woody fruit species, were produced using one of these methods (Parisi *et al.*, 2016). Often the ability to obtain tree saplings with new traits or mutations by genetic engineering depends on the existence of a well-established in vitro regeneration protocol, which depends on the genotype and the type of starting plant tissue used. it is more advisable from an agronomic point of view to in vitro regenerate a new tree from mature tissues, due to the high degree of heterozygosity in majority of the tree species (Cervera *et al.*, 1998; Pérez-Jiménez et al., 2012). In this context relevant progress has been made during the last two decades for some difficult-to-transform woody species, such as peach or grapevine genotypes, in which efficient protocols for the regeneration of adventitious shoots have been developed starting from adult tissues (Sabbadini *et al.*, 2015). Introduction of one or more new genes or regulatory elements using genetic engineering techniques, directly manipulates the genome of an organism in order to express or silence specific traits (Mittler and Blumwald, 2010; Rai and Shekhawat, 2014). New biotechnological tools used for modifying an existing DNA sequence in a plant, comprise of insertion/ deletion and gene replacement, or stable silencing of a gene or promoter sequence. In this category we consider techniques such as RNA interference (RNAi), cisgenesis/intragenesis, trans-grafting, and gene editing techniques including zinc finger nucleases (ZFNs) as well as clustered regularly interspaced short palindromic repeats/CRISPR-associated protein 9 (CRISPR/Cas9 system), to introduce new traits into a host plant genome (Sweet *et al.*, 2017).

Nevertheless, there are no specific investigations of genetic engineering in *Pongamia,* due to limitations of non-availability of basic genetic data. Notwithstanding the gaps in the knowledge, the above mentioned areas of research should stimulate speedy and focussed research in *Pongamia*.

Molecular biological techniques to identify the clones of plants

Commercial plantations in any woody plant species including *Pongamia pinnata* could succeed in terms of the economic viability only with the development of dependable clonal plantation materials of elite genotypes. Nevertheless, till now there are no organized plantations of *Pongamia* due to the paucity of genetically elite clonal planting materials.

A beginning has, however, been made in this context (Prasad, 2013) in *Pongamia.* On the other hand, research in progress in Australia has resulted in productive plantations based on improved genetic materials.

There arises a need to evolve technologies to identify, validate and characterise the superior clones. Some molecular biological techniques are becoming available to meet this need as described below.

- Molecular markers, using primers designed on the basis of DNA sequences of desirable genes, has been developed and validated for clonal progenies of *Leymus* grass hybrid by ARS scientists. (ARS/USDA, 2005), which can be followed and adopted to identify and characterise the clones
- Recent progress in isolating DNA not only from living tissue but also from wood and wood products offers new opportunities to test the declared origin and genetic nature of material such as clonal saplings for plantation establishment or timber (Finkeldey *et al*., 2010). However, since most forest tree populations are weakly differentiated, the identification of genetic markers to differentiate among spatially isolated populations is often difficult and time consuming. Two important fields of "forensic" applications are described: Molecular tools are applied to test the declared origin of forest reproductive material used for plantation establishment and of internationally traded timber and wood products. These applications are illustrated taking examples from Germany, where mechanisms have been developed to improve the control of the trade with forest seeds, seedlings, grafted cones, and from the trade with wood of the important Southeast Asian tree family Dipterocarpaceae. This technology exhibits its validity and is applicable in the case of other woody trees including *Pongamia.*
- The shoot apex of higher plants contains undifferentiated meristematic cells that serve as the origin of post-embryonic organs. The transition from vegetative to reproductive growth results in the commitment of the apical

meristem to floral organ formation. To identify the molecular signals that initiate floral development which are genotypic specific, there is the need to pursue the isolation of genes that are transcriptionally active in the shoot apex of the clone during the transition from vegetative to floral growth. The small size of the apex might lead to utilizing polymerase chain reaction of shoot apices. This approach enables the isolation of the apex-specific and floral apex-specific cDNA clones (Kelly *et al.*, 1990). Each of the clones could exhibit a unique transcript that increases in abundance in the shoot apex during the transition to flowering, showing high levels of expression in developing petals, stamens, and pistils, which could be a valid indication of the genetic nature of the clone.

- Recent advances in our understanding of signal transduction pathways for environmental responses in herbaceous plants, including the identification and cloning of genes for proteins involved in signal transduction, should provide useful leads to unravel the genotypic differences among the clones and varieties of woody plants including *Pongamia* (Jain *et al.,* 2000).

13. SELECTION OF ELITE TREES

Success of *Pongamia* based biofuel programme invariably depends on the choice of elite planting material across the highly populated provenances of central and eastern dry zones in India (Pavithra *et al.*, 2015).

Based on the experience gained in the genetic improvement of oil bearing trees and cashew (Prasad 1994, Prasad *et al.*, 2000), certain attributes of some crucial canopy characters of consistency and stability of expression linked to the kernel yield were employed (Prasad 2013) in the development of basic criteria for selection of the elite trees of *Pongamia* (Fig.13). The tree selection is not based on total pod yield alone, which is highly variable.

The characters chosen are as follows

1. Appropriate canopy architecture with extended globose canopy
2. Intensive branching pattern with higher proportion of pod bearing branches.
3. Cluster bearing habit viz., (i) higher number of clusters (>10) per branch & (ii) more than three pods per cluster (Fig.12).
4. Higher shelling percentage (>40%)
5. Resistance to pod-malformation and
6. Higher kernel yield per tree

The elite trees of *Pongamia* possessing significantly higher levels of kernel yield of the range of over 45 kg / tree were selected based on the above mentioned

Fig. 12. Pod clusters of an elite tree of *Pongamia pinnata*

Fig. 13. An elite tree of *Pongamia pinnata*

stable canopy attributes strongly associated with seed yield. The yield data collected over two seasons (2010-2011 and 2011-2012) demonstrated the consistently higher yield levels of the selected trees (Prasad 2013). The seed oil content of the selected trees ranged from 33 to 39% as against oil content of 28 to 32% in normal and unselected trees (CFTRI 2012) A certain number of elite *Pongamia* trees exhibited higher contents of oil and karanjin, a flurano flavanol of industrial value (CFTRI 2012). Varaprasad (2014) and Varaprasad and Sudhakara Babu (2015) have identified certain improved accessions of *Pongamia* which are presented in Table 8.

Table 8. Improved accessions of *Pongamia* identified in different regions of India (Varaprasad, 2014; Varaprasad and Sudhakara Babu, 2015)

Agroclimatic zone	Improved varieties accessions
Eastern Plateaus & Hills Region	Baraiminda, Mane lies war
Southern Plateau & Hills Region	KNR1-(29.31%), KSRl (27.34%) and TCR-MD (27.2 %), TCR-Cl-1, PT 1 and PT 2 of Dharwad.
Central Plateau & Hills Region	NRCP-13 (early flowering), NRCP 24, NRCP 26 (high yielding), IC 527952 (34 %)
Eastern Plateaus & Hills Region	KKYPP-02 (31.7%), KKVPP-17 (42%)
Central Plateau & Hills Region	RPK 14
Western Plateau & Hills Region	RHRAK 50, RHRAK 44

% Oil content in parenthesis

14. PLANT PHYSIOLOGY

One of the problems faced in raising plantation of *Pongamia* is its low seed viability. The seeds from freshly harvested well dried pods show a germination percentage of 70 to 85; but the seed viability was observed to decline after 4 to 5 months of shelling from freshly harvested pods. Kumar *et al*., (2011) conducted an experiment to elucidate the possible physiological and biochemical changes associated with seed deterioration during storage of karanj (*Pongamia pinnata* L.) seed. The seeds of variety 'Bak 49' ('IC430529') were extracted, processed and stored at three different levels of relative humidity (RH) (75%, 33% and 5.5%) and temperatures (4°C, 20°C and ambient). The seed viability pattern, physiological and biochemical parameters under different conditions were monitored at regular intervals to assess the effect of storage. Among the different RH treatments, 33% RH showed significantly higher values for viability and vigour over 5.5% and 75% RH at all temperatures. The biochemical parameters like electrical conductivity of seed leachates and lipid peroxidation under different treatments showed significantly increased values with seed deterioration. The level of total soluble sugars increased gradually, whereas total soluble proteins and enzyme activity (dehydrogenase and acid phosphatase) decreased with storage period in all the treatments. The optimal conditions for extending seed storability in karanj without having any adverse effect on physiological and biochemical parameters were 4°C and 20°C and 33% RH. This study could possibly be helpful for conventional storage of pongamia seeds on large scale and can be further exploited in other orthodox tree species.

In the flowers of *Pongamia pinnata* only one of the two ovules develops into a seed in most of the pods. Since pollen was not found to be limiting, reduced fertilization could not completely explain the observed frequency of seed abortion. It probably implies an effect of post-fertilization factors. The studies carried out by Arathi *et al.*, (1999) using aqueous extracts of developing seeds and maternal tissue (placenta) did not influence abortion in vitro, suggesting that abortion may not be mediated by a chemical. In their studies involving the experimental uptake of 14C sucrose *in vitro* indicated that both the stigmatic and the peduncular seeds have similar inherent capacities of drawing resources; but the peduncular seed is deprived of resources in the presence of the stigmatic seed. According to them this deprivation of the peduncular seed could be offset by supplying an excess of hormones leading to the subsequent formation of two seeds in a pod. The prevalence of single-seeded pods in *P. pinnata* seems therefore to be a result of competition between the two seeds for maternal resources. This phenomenon perhaps is related to the evolutionary implications with respect to possible dispersal advantage enjoyed by such pods.

A compound viz., Placobutrazol [*4- chlorophenyl dimethyl triazol pentan*] sprayed @ 10 g of active ingredient per tree twice in a year was fond to induce inter-node compression resulting in height reduction leading to a more compact canopy with darker green foliage with higher photosynthetic efficiency and delayed leaf senescence (CRIDA 2012). It was also noticed that all the mentioned favourable changes conferred resistance to environmental stress in *Pongamia*. The major limitation in this context, however, is prohibitively high cost of the compound in question.

15. CARBON SEQUESTRATION BY *PONGAMIA*

The trees sequester carbon dioxide by storing it in their trunks, branches, leaves and roots; the best trees for carbon dioxide absorption will have dense wooded canopies.

The Carbon sequestration potential of *Pongamia pinnata* during the 10 to 15 years of its growth was found to be many folds higher than that of several other tree species. *Pongamia* was found to sequester around 45 to 50 kg of C per tree per annum as against 28 to 35 kg of Neem (*Azadirachta indica),* 23 to 26 kg of Mahua (*Madhuca latifolia)* and 11 to 15 kg in respect of Tendu (*Diospyros melanoxylon*). (Table 10.) (Chaturvedi *et al*., 2011; Jesse More, 2009; Gupta, 2009; Kamarkar *et al*., 2012; Gera and Chauhan, 2010 and Reddy *et al.,* 2009) (Table 9).

Table 9. Carbon sequestration by some popular tree species.

Tree type	C sequestration / tree of 15yrs (kg)/ annum	C Sequestration (t)/ ha	Tree density (No./ha)
Pongamia (*Pongamia pinnata)*	45-50	23.5	500
Tendu(*Diospyros melanoxylon)*	11-15	6.0	460
Mahua (*Madhuca latifolia)*	23-26	8.5	354
Neem (*Azadirachta indica)*	28-35	13.0	450

A conservative estimate of carbon sequestration by *Pongamia i*s as follows: A population of 200 trees (aged 20 years) in an area of one acre can sequester 8,200 lbs or 6118.2 kg of Carbon (VAYUGRID, 2014). Some estimates arrived at by CSIR indicate that the above figure can go up to 25 tons /ha.

A study carried out in 2006 estimated that over the course of a 25-year period, one *Pongamia* tree has the potential to sequester 767 kg of carbon. The carbon sequestration ability of pongamia was calculated for 3,600 trees planted in the Powerguda village in Adilabad district of Telangana State in India. The certified carbon emission reduction was sold to '500ppm GmbH', which is an environmental group of Germany. The reduction came from replacing diesel fuel with natural oil from *Pongamia* seeds (Dsilva, 2015). The purchase was effected for ten years' supply of emission reduction from 140,000 kg of *Pongamia* oil, worth $4,164.

According to a USA based Company TerViva (Neo, 2020), over a span of thirty years of its life, a single acre of *Pongamia* can sequester up 115 tons of Carbon and one acre of *Pongamia* can replace one acre of oil palm plantation or up to 4 acres of soybean. Accordingly, the food products developed out of Pongamia by TerViva would be Carbon negative since the trees sequester more Carbon than the beans that produce in emissions from field to plate (Neo, 2020).

The current trends of global warming are of great concern. It is estimated that an increase of 2.5% of global temperature would reduce agricultural productivity in USA by 6%; but by 38% in India (Sengupta, 2013). It is therefore necessary to come up with appropriate technologies involving tree species that are efficient in cutting down global warming and bring about perceptible levels of C sequestration in addition to its significant oil yield, which renders it to form a good feed stock for bio-deisel production. *Pongamia* is one such species that need attention in this regard. However, it is necessary that such tree types grown for bio-fuels, must not occupy fertile irrigated lands to avoid competition with regular food and commercial crops. *Pongamia pinnata* meets the above condition by virtue of its drought resistance and hardy nature and as such forms an excellent feed stock for bio-fuel production.

16. PHYTOREMEDIATION OF PROBLEM SOILS USING *PONGAMIA*

Phytoremediation is the direct use of living green plants *in situ* for removal, degradation or containment of contaminants in soils, sludge, sediments, surface water and ground water. Phytoremediation is a cost effective, solar energy driven clean-up tcchniquc.

Fig. 14. Three-year-old elite *Pongamia* plantation on copper mine toxic dumps in Zambia (Courtesy: Dr Benjamin Warr, Zambia) (Warr, 2018.)

Pongamia has been reported to be efficient in remediating problem soils containing heavy metals and other pollutants that impede plant growth (Prasad 2007, Tulod *et al.,* 2012 and Kumar *et al.,* 2017). The phytoremediation potential of *Karanj* has been examined in few comparative studies; in pot experiments, on coal mine spoil, fly-ash amended soils and on landfill wastes (Pandey 2017).

Total concentrations of copper in growing media declined as a result of plant uptake (Shirbate and Malode 2012). The addition of Vesicular-Arbuscular Mycorrhiza (VAM) and zeolite improved growth and therefore total uptake (Tulod et al 2012).

Selective compartmentalisation indicates the highest concentrations of copper are in roots, seed coat, leaves, shoots (1220 ppm to 26 ppm respectively) (Kumar et al., 2009). The on-going *Pongamia* plantation in Zambia planted during November 2016 (Fig. 14) is the first to provide evidence of phytoremediation by virtue of excellent tree survival and seed production on disused copper tailings facilities (Warr 2018).

Warr (2018) has shown how genetically elite *Pongamia* (*Karanj)* could be used reclaim mine overburdens of copper mine zones in Zambia and render them suitable for crop growth.

Following are the mechanisms involved in phytoremediation (Etim 2012)

Phytoextraction is also called phyto-accumulation, refers to the uptake and translocation of metal and chemical contaminants in the soil by plant roots in the above ground portions of the plants. Certain plants, called hyper-accumulators, absorb unusually large amounts of metals in comparison to other plants.

Rhizofiltration is the adsorption of soil and groundwater contaminants onto, or into, plant roots. Rhizofiltration is similar to phytoextraction, but the plants are used primarily to address contaminated groundwater rather than soil.

Phytostabilization is the use of certain plant species to immobilize contaminants in the soil and groundwater through absorption and accumulation by roots, or precipitation of the contaminants within the root zone of the plants. This process reduces the mobility of the contaminants and prevents migration into the groundwater or air and reduces the bioavailability and thus entry into the food chain.

Phytovolatilization is the uptake and transpiration of a contaminant by a plant, with release of the contaminant, or a modified form of the contaminant, into the atmosphere.

Phytodegradation: This is also referred to as phyto-transformation. It involves the degradation of complex organic molecules to simple molecules or the incorporation of these molecules into plant tissues. When the phytodegradation mechanism is at work, contaminants are broken down after they have been taken up by the plant. As with phytoextraction and phytovolatilization, plant uptake gener generally occurs only when the contaminants' solubility and hydrophobicity fall into a certain acceptable range.

Rhizodegradation refers to the breakdown of contaminants within the plant rhizosphere. It is believed to be carried out by bacteria or other microorganisms whose numbers typically flourish in the rhizosphere. Microorganisms may be so prevalent in the rhizosphere because the plant exudes sugars, amino acids, enzymes, and other compounds that can stimulate bacterial growth. The roots also provide additional surface area for microbes to grow on and a pathway for oxygen transfer from the environment. The localized nature of rhizodegradation means that it is primarily useful in contaminated soil, and it has been investigated and found to have at least some successes in treating a wide variety of mostly organic chemicals, including petroleum hydrocarbons, polycyclic aromatic hydrocarbons (PAHs), chlorinated solvents, pesticides, polychlorinated biphenyls (PCBs), benzene, toluene, ethylbenzene, and xylenes (EPA, 2000). It can also be seen as plant-assisted bioremediation, the stimulation of microbial and fungal degradation by release of exudates/enzymes into the root zone (Zhuang *et al.*, 2005).

17. SYMBIOTIC NITROGEN FIXATION, SOIL AMELIORATION and ROOT PENETRATION IN ROCKY TERRAIN BY *PONGAMIA*

Pongamia produces its own nitrogen through Symbiotic Nitrogen Fixation, thereby displacing approx $200 / ha/year of nitrates applied as compound fertiliser. Soils under *Pongamia* stands of +15 years' age accumulate almost 1100 to 1300 kg of Nitrogen per hectare per annum, apart from enhancing the soil organic matter and consequent positive soil biological activity (Elevitch and Wilkinson, 1999).

Table 10. Data on analysis of soil under *Pongamia* and of adjoining Barren soil (Vayugrid, 2014)

Parameter	Under Pongamia (6yr old trees)	Barren soil
pH	6.5	6.7
Bulk density (g/cubic cm)	4.75	4.87
Oxidisable C (mg/g of soil)	1.6	0.45
Soil organic matter (%)	3.27	1.16
N (%)	0.091	0.027
P (%)	0.0093	0.0030
K (%)	0.086	0.032

The soil under six-year-old trees of *Pongamia* exhibited marked improvements not only soil organic matter and Nitrogen, but also in other plant nutrients viz., Phosphorous and Potash. (Table10)

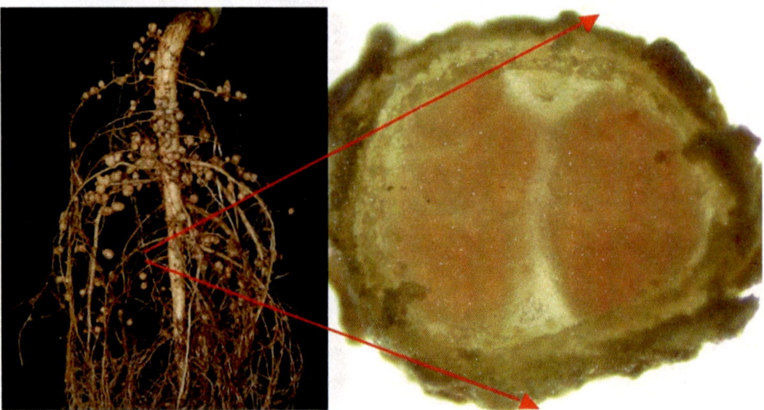

Fig. 15. Rhizobial root nodules of *Pongamia pinnata* with section of a magnified root nodule showing rhizobial activity. (Source: Biswas *et al.*, 2013).

According to Samuel et al (2013) nodule number in legumes is controlled by a systemic regulation loop called AON (Autoregulation of Nodulation), in which nodule primordia signal the leaf to produce an inhibitor blocking further meristem proliferation. Thus nodules are usually found on the upper regions of a legume's root system.

Root penetration through rocky terrain

Pongamia tree can grow even on rocky terrains. The studies showed that about one-quarter to one-third of the total length of the root length could be in the rocky layer. It has been observed that the tree root penetration is assisted largely due to its symbiotic association with mycorrhizal fungi. Arbuscular mycorrhizas, or AM (formerly known as vesicular-arbuscular mycorrhizas, or VAM), are mycorrhizae, whose hyphae penetrate plant cells, producing structures that are either balloon-like (vesicles) or dichotomously branching invaginations (arbuscules) as a means of nutrient exchange. Mycorrhizal infection occurs only on the smallest order of secondary roots. These are the root-tips that are still growing, elongating and increasing in girth, since this is the only part of the root system that will absorb water and minerals. In all arbuscular mycorrhizas fungal (AMF) infections, only the cortical cells of the root are invaded by the fungus. This is the area of the root between the epidermis and the vascular tissue. This phenomenon is a finely balanced symbiotic relationship in which the fungus gets the nutritional carbohydrates from the tree, while it provides the host with the mineral nutrients to roots at the same time aiding the root penetration through rocky terrain. (https://www.sciweb.org/science2)

18. GREEN COAL FROM *PONGAMIA*

Pod shells of *Pongamia pinnata (Karanj)* could be converted into briquettes (green coal) to be used as an efficient fuel in rural homes and as source of energy in rural zones. Density of the Briquettes is estimated to be 0.6819 g/ml with calorific value of 4052 cal/g. Each tree can produce about 0.1 kwh from the pod shells each year, which comes to almost 6kwh over its life time.

The data presented in Table 11 show the efficiency of the green coal as compared to the traditional black coal.

Table 11. A comparison of black coal and green coal

	Black Coal	Green Coal
	3800 - 5000	4500
Sulfur Content	0.5% - 1%	0%
Ash Content	35% - 45%	~0%

(*Source:* Vayugrid, 2014)

19. PROPAGATION

Pongamia can be propagated by seed as well as by vegetative means. The techniques of propagation and their evaluation in relation to the improvement of *Pongamia pinnata* have been well reviewed by Mukta and Sreevalli (2010).

i) **Seed propagation** is widely adopted in afforestation programmes due to ease and feasibility (Allen and Allen 1981, Duke 981 and1983). In order to obtain healthy and good quality seeds, their collection from selected mother trees, extraction and storage in a scientiûc manner needs attention. Therefore, principles for seed-testing for the determination of purity, germination, moisture content and freedom from pathogens according to rules laid down by ISTA (Anonymous,1966) need to be applied. Seeds lose their viability with in a period of six to eight months if not stored properly under optimum temperature conditions. Seed viability, however, may be enhanced to one year following harvesting, if stored under temperatures lower than 20°C. (Rout-Sandeep and Naik-Saswat, 2015) The ideal storage conditions are always those which reduce the growth processes of respiration and transpiration to the lowest possible degree, without impairing the inherent vitality and strength of the seed embryo (Phartyal *et al.*, 2002). Seeds of *P.pinnata* stored in the middle layer of glass-stoppered containers retained higher viability even after 1 year and exhibited higher germination rate (Athaya, 1985). Also Seed viability may be better ensured if the seeds are stored as whole unshelled, sun-dried, uninfected, bold and healthy pods.

Seed germination could be enhanced by steeping the fresh seeds of *Pongamia* in hot water of around 50°C for 20 minutes before sowing. Seeds are sown in seed beds/polythene bags/sand trays with the micro-pile facing downwards not deeper than 3 cm to the top of seed and 5 cm to the bottom. Mixture consisting of 1part of sand,1 part of red soil and 1 part of well decomposed and grounded cattle manure has been found to yield good results for filling-up of polythene nursery bags and preparation of seed beds for sowing. Seed germinates within two weeks of sowing.

Seed propagated tree populations exhibit a wide variation both in terms of growth and development, branching, seed and pod and seed characters. Some seeds throw out seedlings with less or no chlorophyll termed as *albina, xantha, chlorina* etc., if the seeds happen to be derived from self-pollination. Raising of healthy seedlings in well managed nursery is absolutely necessary for the production of healthy and vigorous root-stocks for grafting.

Transplanting the seedlings into the field should occur at the beginning of the rainy season when seedlings are 60 cm in height with well-developed root system. Soil mass should be retained around the roots during transplanting.

ii) **Propagation by stem cuttings** is successful and is practised to certain extent (Palanisami *et al.,* 1998). Stem cuttings, however, result in fibrous root system of the saplings, which renders them vulnerable to drought under rainfed conditions. Also, the fibrous root system does not offer good anchorage to the growing sapling under fierce windy conditions which are very common.

   Additionally, with stem cuttings it would not be feasible to produce vast planting materials for raising plantations in large areas. Despite the limitations associated with the stem cutting propagation, the ease of doing it encourages many to resort to this process. In such cases, adhering to the following recommendations (Mukta and Sreevalli, 2010) would be useful.

- Indole butyric acid (IBA) at 1000ppm as an alcoholic dip induces best rooting in cuttings.
- Cuttings root better in the months when cambium is in active phase.

iii) **Root suckers** are rather plentiful as well. It is a rapid-growing coppice species that can be cloned.

iv) **Pongamia can also be propagated by grafting**, by far the best method of propagating valuable genotypes and high yielding mother trees.

v) **Tissue culture technology for propagation of elite trees**: The work pertaining to tissue culture of *Pongamia* is included under the topic Cytology and Genetic improvement. A critical examination of the work carried out in this area indicates some success in regeneration & micro-propagation (Sujata and Harzra 2006, Sugla *et al.,* 2007, Sujatha *et al.,* 2008 and Biswas *et al.,* 2013). However, micro-propagation technology has not yet come to the stage of developing plantlets/clones for commercial plantations on large scale.

One of the major limitations associated the tissue cultured plants of *Pongamia* are week root system that can't withstand the rigours of dryland rain grown conditions (Vayugrid, 2014).

vi) Production of Elite clones through grafting technology: The grafting techniques involve the use of root stock seedling over which the vegetative scions are grafted by several methods viz., tip grafting, veneer grafting, bud-grafting or budding and wedge-cleft grafting etc., which are normally employed in the vegetative propagation of trees (USAID 2007). Cleft-wedge grafting has been found to be most successful (95%)using 3-month-old seedlings raised in poly-bags as the stock and semi-hardwood scions of 12–15cm length (Karoshi and Hegde, 2002).

However, it has been observed that higher rate of success could be achieved in *Pongamia* using root-stocks with soft-wood (75 to 80 days old vigorous root-stocks) employing cleft-wedge grafting (Fig.16); which has been modified and standardized for vegetative propagation of cashew (Prasad *et al.,* 2000, NRC 1994).

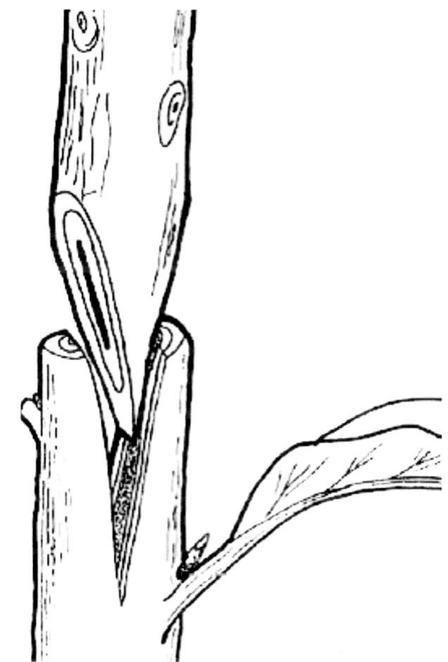

Fig. 16. Cleft-wedge grafting technique

Under the aegis of Vayugrid, grafted saplings were developed by adopting a special grafting technology consisting of appropriate root-stock age and vigour resulting in a true copy of its genetically elite mother *Pongamia* tree. The following steps are involved in the development of elite grafted saplings of *Pongamia pinnata* (Prasad 2013)

1. Identification and selection of genetically elite *Pongamia* trees based on the stable selection criteria developed during 2009 to 2016 in Vayugrid.
2. Growing of vigorous root-stocks adopting appropriate nursery technologies,
3. Grafting dormant scion stem tissues of the elite mother trees on to the root-stocks possessing soft wood, employing a special and efficient grafting technology and

4. Nurturing of the fresh grafts to a suitable stage of development in the nursery (Fig. 17) for their transplantation in the main filed during the rainy season.

Direct planting of seeds in the field to raise root-stocks and *in situ* grafting with a scion from high yielding trees after 9 months has also suggested (Wani and Sreedevi, 2005); but is found to be beset with problems (Vayugrid, 2014)

Fig. 17. Pongamia Nursery with elite grafted sapling

20. PRODUCTION TECHNOLOGY FOR ELITE *PONGAMIA* PLANTATION

Till recently *Pongamia* has been employed extensively by the Forest Department for afforestation. In the farming sector *Pongamia* receives no importance, notwithstanding its valuable products and several beneficial attributes. *Pongamia* trees are usually originated from self-sown seeds and the resulting trees don't receive any attention by the farmers.

The concept of growing *Pongamia* in an organized plantation has not taken roots in India. Hence the differences between *Pongamia* planted for afforestation and proposed elite *Pongamia* plantations have to be understood and appreciated. In the case of *Pongamia* used for afforestation, the tree becomes a small component of the complex tree populations used for this activity. *Pongamia* trees planted in the above kind of programmes can't be managed towards achieving sustainable economic yield levels.

In the case of elite *Pongamia* plantation on the other hand, high quality genetically elite grafts of only one species viz., *Pongamia pinnata* form the planting materials and the plantation is raised in a definite plantation geometry and population density and managed scientifically to achieve the economically sustainable levels of productivity.

Currently 3% of the world's forests are plantations, consisting of 60 million hectares in developed nations and 55 million hectares in developing nations (WRI,1998; FAO, 1999).

In some regions, plantations comprise a major proportion of forest area, including 44% in Japan, 20% in New Zealand, and over 90% in Britain (Donald *et al.*, 1997; FAO, 1999). Though tropical forest cover is declining throughout the world, tropical plantation area has increased dramatically, from about 10 million hectares in 1980 to about 44 million hectares in 1990 (Lugo, 1997). Plantations can realize much more production of the economic produce on a much smaller land base (Karen *et al.*, 2010, JudyLoo *et al.*, 2014).

Indeed, some calculations indicate that plantations could meet the demand of most industrial products demands on <10% of the world's forest land area (Sedjo and Botkin, 1997).

Establishment of an elite *Pongamia* based bio-energy plantation is a long term investment and deserves a very critical planning. The selection of proper location and site, planting system and planting distance, choosing the planting materials have to be considered carefully to ensure maximum production. The following plantation management technologies have been developed by the author under the aegis of Vayugrid (Vayugrid, 2014).

Certain plantation management strategies specific to *Pongamia* are as follows:

- Maintain diversity of genotypes: Grow clones from diverse elite trees rather than those from one or two mother trees, if there is no consideration of a specific product from plantation
- Eco-friendly management: Avoid use of chemical pesticides and other strong chemicals that would harm the symbiotic biological systems including insect pollinators.
- Ensure soil conservation avoiding erosion and carry out water harvesting by collecting run-off in farm ponds exploiting the natural slope.
- Nutritional management must be effected through mostly organic sources.

i) Planting and initial care of planted grafted saplings

The investigations carried out at the ICAR-Central Institute of Dryland Agriculture, Hyderabad have come out with the following recommendations.

- It is recommended that pits of the size of 45-50 cubic cm be dug before the rainy season.

- Under semiarid conditions and on light soils of low fertility, a spacing of 5m X 4m is recommended for planting, leading to a density of 500 trees per hectare.
- The excavated top soil is mixed with sand and organic manure / FYM in the proportion 1 (sand): 2 (soil): 1 (FYM) and 100 g of neem cake with which the pits are filled at the time of planting with the receipt of rainfall of at least 50 to 60mm or before surface soil moisture gets exhausted.
- It is desirable to make a basin at the base of each planted sapling to conserve rainwater and support the sapling with a stake till it gets established. In order to keep the seedling stable, the soil around it should be pressed by hand and the seedling covered with grass straw after planting.
- Weeding may be carried out around each sapling after 60 days of panting and till the plant gets established with firm and deep root-system in the soil
- No irrigation may be needed normally as *Pongamia* saplings can withstand even moderate to severe soil drought.
- During the first summer/rainless hot period following planting, watering of saplings with 30 to 40 litres of water/plant either from water harvested from run-off or other source would ensure proper canopy development.

In areas with exposed rock, artificial explosion can be used for pit preparation for planting high value elite grafts. Powder is used to explode a pit to a depth of 1 m and a diameter of 2-3 m. Then good top soil and organic matter are put into the pit into which an elite graft is planted and watered. The nitrogen from the explosion powder remains in the pit after explosion thus increasing the nutrients for tree growth.

Following two important steps should be implemented to ensure proper seedling establishment and zero plant mortality in the first year of the plantation.

- Undertake planting of elite grafts only when the soil profile is saturated with rain water to a depth of more than 60 cm.
- Mix the dugout soil with 100 g of neem cake and fill the pit at planting to ward of damage to root by any soil born biotic agents.

ii) Nutrient Management

As flowering season approaches in October to December, trees need to be replenished with nutrients in order to have enough food for the developing pods. The decision for supplementary nutrition may be taken depending upon the growth and development of canopy. If the growth of sapling is stunted after the first year, the following procedure may be followed. For this a small trench of 30 cm deep, 120 to 150 cm long and 40 to 60 cm wide should be made around the tree canopy. Around 2 kg of FYM + 50 g of SSP may be applied in the trench dug around the plant and the trench closed with the dugout top soil.

Since *Pongamia* plantations are raised on soils of low fertility and also impoverished soils, it is possible that the elite grafts might exhibit some nutritional deficiencies. If the symptoms of micronutrient deficiency are observed, the following corrective measures must be taken. Soil and leaf analyses are frequently used complementary tools for decision-making.

Micronutrient Trace elements such as magnesium, iron, zinc, manganese, copper, boron, molybdenum, etc. are rarely applied systematically.

These elements are sprayed on to fully developed young leaves of the spring growth, if leaf analysis results show the need.

MAGNESIUM: Signs of this deficiency are found only in adult leaves.

Discoloration starts in the median part of the lamina and spreads gradually to the base and tip of the leaf in a V-shaped pattern. In the final stage, only a V-shaped part close to the petiole is not discoloured. The young leaves of trees suffering from this deficiency are never affected. Spray new leaves with 1.2% magnesium nitrate on completion of lamina elongation.

IRON: The deficiency is shown by overall discoloration of the lamina, which turns pale green to yellow. In contrast with nitrogen deficiency, the vein pattern remains green.

True soil iron deficiency is rare. It is more frequently a blockage of assimilable forms of iron. If possible, first correct the physical causes of the deficiency viz., hydromorphy, alkaline pH, etc. Apply chelated forms of iron to the soil.

ZINC: This is a frequent deficiency in all areas, seen in the young leaves which remain small and erect. The lamina is seriously discoloured-from pale green to yellow-in all the inter vein parts. The veins and halo along each of them remain green.

Spraying new leaves with zinc sulphate (2.5 g/l) on completion of lamina elongation.

MANGANESE: This is another common deficiency. It is similar to zinc deficiency but differs in two respects: the leaves are not deformed and remain their normal size and lamina discoloration is less intense. Spraying new leaves completing lamina elongation with manganese sulphate (1.5 g/l).

BORON: This fairly rare deficiency is shown by the suberisation of the veins on the underside of the lamina. The young shoots may wither.

The distance between deficiency and toxicity is small and so boron applications must be carefully measured. Spraying with 2.5 g/l Solubor.

Despite lack of any clear data regarding the yield losses in *Pongamia* due to micronutrient deficiencies, symptoms of deficiencies of certain micronutrients, notably that of zinc and Magnesium have been observed by the author at different stages of plant growth, particularly when no organic matter was applied at planting. Application of organic matter in the form of well decomposed FYM @ 4kg/ plant or 2 kg of poultry manure /plant or 1.5 kg of Organic Matter (Annapurna of Multiplex) would be necessary to ward of any micronutrient deficiencies.

iii) Biotic Stresses and Plant Protection

Although *Pongamia pinnata* attracts many pests and diseases, it is unique in the sense that the biotic stresses have not caused much of yield reduction. Some of the important pests are *Parnara mathias, Gracillaria* spp, *Indarbela quadrinotata, Myllocerus curvicornis, and Acrocercops* spp. Attacks by these insects cause whitish streaks and the formation on Pongamia of galls on affected leaves.

Fig. 18. Leaf galls induced by the mite *Aceria pongamiae* on pongam leaves

Eriophyd mite (*Aceria Pongamiae*) is known to induce leaf galls in India (Fig.18). *Aceria pongamiae* is a highly host specific eriophyid mite, producing varying numbers of small finger-like leaf galls on *P. pinnata*. The number of galls on infested leaf varies, quite often individual galls fused to form complex, irregularly shaped, massive structures, covering entire laminar area including the midrib, vein and vein lets. Each gall carries hundreds of mites in different stages of development, namely, the egg, 1st nymph, 1st quiescent stage, 2nd nymph, 2nd quiescent stage and adult male and female\ on the leaves of *P.pinnata.* (Nasareen et al., 2013). No plant protection measure is recommended as the mite induced leaf galls are not known to adversely affect the pod and seed yield.

Another insect pest of significance is a dipteran fly viz., *Aspondylia pongamiae* (Fig.19) that causes damage to the fertilized ovary, transforming pods into seedless rounded sterile structures. The laraval infestation of flower directly affects the production of seeds. Detailed studies were conducted (Devaraj and Sundarraj 2014) from 2007 till 2009 to identify the natural parasitoids of *A. pongamiae*. The study revealed the occurrence of four species of hymenopteran parasitoids, namely, *Eurytoma dentata, Megastigmus albizziae, Neanastatus proximus and Ormyrus kama*. The presence of all these

Fig. 19. Adult fly of *Aspondylia pongamiae*

parasitoids on *A. pongamiae* formed first records. Among these parasitoids maximum parasitisation was by *E.dendata* followed by *M. albizziae, N. proximus and O. kama*. Combined parasitisation, amounting to 37.56% in 2008-2009 and 44.54% in 2007-2008 was observed indicating that these parasitoids played significant role in keeping the population of *A. pongamiae* under control.

Several fungi attack the seedlings and the trees. *Ganoderma lucidum* causes root rot and *Fomes merilli (Murrill) Sacc. & Trott* attack the tender shoots and leaves and cause early defoliation in the seedlings and trees.
Pongamia leaf miner has not been found to cause any yield reduction.

Phytophthora gummosis is certainly the world's most widespread fungal disease.

Three *Phytophthora* species are involved in particular: *P. parasitica, P. citrophthora* and *P. palmivora.* The pathogenic fungus is endemic in the soil in all zones. It is particularly active during hot, humid weather. *Phytophthora* can survive in a suspended form during dry periods that are unfavourable for its development. In citrus, attacks may occur at all stages of cultivation-on both young seedlings and adult trees in orchards. In the parts of the plants attacked, the fungus is found in living tissue close to the lesions. The attacks destroy the bark on trunks and low branches; this strongly disturbs the translocation of elaborate sap and causes flows of gum and the yellowing of foliage upstream of the area of bark that has been destroyed. Other less visible symptoms include root rot.

Unless trunk and root attacks are controlled at an early stage they will cause the total or partial withering of the tree. This disease however is rare in *Pongamia.* Also *Pongamia* is found to be resistant to termite attack.

In order to achieve a regular plant protection-cover conducive for pest and disease avoidance, without any heavy investment on chemicals, it is recommended that a routine sprays of the following organic fluid products be carried out twice every year starting from the third year particularly at (1) pre-flowering and (2) peak flowering phases.

Fig. 20. *Dendrophthoe falcata* parasitic plant on one tree of *Pongamia pinnata* near Hyderabad Oct- 2019

Locally available organic products recommended for *Pongamia* plant protection

- Neem oil spray / neem leaf extract
- Custard apple leaf extract
- Coconut leaf extract
- *Pongamia* leaf extract

A Phanerogamic parasite viz., *Dendrophthoe*

falcata (L.f) Ett. belonging to the family Loranthaceae (Selvi and Kadamban, 2009) was observed on one tree of *Pongamia pinnata* near Hyderabad for the first time in October 2019 (Prasad, unpublished) in the zone of Telangana (Fig. 20). Selvi and Kadamban (2009) however, reported the occurrence *D. falcata* on a wide range of tree species including *Pongamia pinnata* in southern Tamil Nadu. The parasite if it occurs, should be eliminated and destroyed right at the initial phase with a constant vigil and supervision of the plantation.

iv) **Nipping of root-stock sprouts**: It is necessary to carry out nipping / removal of sprouts arising from the root-stock zone of the grafted sapling. Also eliminate all root sprouts and maintain the principal / main trunk clean devoid of side growth up to a height of 1 meter.

v) **Irrigation**: No artificial irrigation may be necessary, as the grafted saplings are highly drought resistant by virtue of the deep rooted root-stock and as such can thrive well with rainfall moisture. Nevertheless, it should be ensured that the adverse impact of drought is minimised to recover tangible levels of productivity.

   In the event of acute drought due to total or near total failure of seasonal rainfall, periodical sprays of canopy with 2% Urea solution once in ten days or fortnight depending upon the soil moisture level would protect the plantation from yield decline.

vi) **Inter-cropping** of drought resistant annual crops viz., castor, pigeon pea, cowpeas, green gram, sesame, cluster beans, horse gram niger, sunflower, fodder grasses etc., is recommended depending upon the local conditions in inter-row spaces of the plantation in the first three years.

   Inter-cropping may not be possible after three years of establishment of plantation.

Fig. 21. Inter-cropping in *Pongamia* plantation.

vii) **Yield and Harvest:** A normal pongam tree starts bearing pods after the 5th or 6th year, giving an average kernel yield of 2 to 3 kg per tree. A genetically superior grafted sapling will start yielding around 3 to 5 kg of kernels per tree after the third to fourth year of planting. The yield increases progressively with age of the tree up to the 25th to 30th year (Table 12) after which the yield may not increase (VAYUGRID 2014).

Mechanical harvesting employing tractor linked mechanical trunk agitator is already in vogue in Australia, which can be implemented in India and other countries too.

Otherwise, periodical manual harvesting of pods may be resorted to. The experience of the author is that most of the elite grafts are synchronous in their bearing and as such number of pod pickings may be reduced to 2 to 3 at harvest.

Table 12. Estimated Pod and oil yield of elite *Pongamia (Karanj)*

Tree Age (yrs) →	3	4	5	6	7	8	9	10	11	12	13	14	15
Pod yield/ tree (kg)	7	15	30	45	58	67	77	89	100	110	120	135	150
Oil yield/ acre (Tons)	0.2	0.4	0.7	1.1	1.5	1.7	1.9	2.2	2.5	2.8	3	3.4	3.7

(*Source:* VAYUGRID 2014)

viii) **Oil Yield and Profit:** Prasad (2019) reported the following oil yields and profits that could accrue form Pongamia production.

- *Pongmia (Karanj)* is grown normally under unirrigated conditions, for which the recommended tree spacing is 5 m X 4 m.
- A tree population of 200 trees is recommended per acre under rain grown dry-land conditions.
- The average oil content under the conditions of dry-land plantation management varies from 35 % to 40%, the average oil content being 37%.
- The mean seed weight is 1.2 to 1.4 g/ seed. However, genotypes of *Pongamia* with 2.0 g of weight per seed are also available.
- The average shelling percentage (pod to kernel ratio) is around 40%.
- The elite grafts of *Karanj* start yielding from the fourth year. The pod / kernel yield goes on increasing from the 5th year, as given the Table 13.
- The average kernel yield from a plantation of one acre of 8 to 15 years' age is around 9 to 10 tons.
- The corresponding oil yield @ 37% will be 3.7 tons /acre
- The bio-diesel yield / ha based on the above calculations will be around 4 (four) tons per acre
- ***Total cost of production/acre from the first year of planting up to the 10th year is around Rs. 70,00/-*** Taking an average for one year; the value will be **Rs. 7,000/- per year per acre**. The major components of cost of production are in the first year towards the cost of the elite saplings (@ a Rs.100/-per plant) and for digging pits and planting operation (@ R.50/- per pit/sapling)
- Profit in terms of kernel sale (@ Rs.30/-per kg) shall be around Rs. 26,000/ - average per year per acre

- Profit in terms of oil yield (price of oil @ Rs.60/- per kg) will be around Rs. 200,000/-per acre per year.
- The profit that would accrue from seed cake and briquettes (from pod shells) would be extra gains, which have not been included in the document.
- Biodiesel: The high cost of vegetable oils would push the biodiesel cost to over $1.00 (around Rs.70/-) per litre. Costs in a small plant would be about 10 US cents per litre higher than the tallow cost. https://www.google.co.in/search?q=Price+of+biodiesel&rlz=1C1E KKP_enIN688IN688&oq=Price +of+biodiesel &aqs=chrome.. 69i5912j69i 57j013.10754 j0j8&sourceid =chrome&ie=UTF-8)
- The current quoted price of biodiesel is around Rs.46/- to Rs. 50/- per litre, which is lower than the market price of *karanj* oil. In view of the above the profits from the production and sale of biodiesel in the case of *karanj* have not been included in this document.

From the above it is evident that the major components of cost of production are in respect of the price of the elite *Pongamia* saplings and the cost of digging pits in the first year of the plantation. The subsequent cost of maintenance of the elite plantation would be very low since the trees do not need any special or expensive nutrient application and irrigation. The cost of plant protection too would be minimal in view of the hardy and pest repellent nature of *Pongamia pinnta,* due to the presence of karnjin and pongmol in the plant system.

Fig. 22. A view of the three and a half years' old young elite clonal plantation of *Pongamia pinnata* on typical dryland of low rainfall zone of Hyderabad

References

Ahmed M, Zafar M, Khan MA, and Sultana S 2009. Biodiesel from *Pongamia pinnata* L. oil.

A promising alternative bioenergy source, Part A: Recovery, Utilization and Environmental effects. **31:** 1436-1442.

Allen ON, and Allen EK 1981The Leguminosae. *The University of Wisconsin Press* 812 pages.

Al Muqarrabun LMR, Ahmat, Ruzaina SAS, Ismail NH, and Sahidin I 2013. Medicinal uses, phytochemistry and pharmacology of *Pongamia pinnata* (L.) Pierre: A review, *Journal of Ethnopharmacology* **150:** 395-420.

Altpeter F, Baisakh N, Beachy R, Bock R, Capell T, Christou P, et al 2005. Particle bombardment and the genetic enhancement of crops: myths and realities. *Mol. Breed.* **15:** 305–327.

Anonymous 1966. International rules for seed testing procedure. *International Seed Testing Association,* **3:** 1–132. Zürichstrasse 508303 Bassersdorf Switzerland.

Anonymous 2002 *The Indian Express* June 02, 2002.

Anonymous 2012. Combined Discussion Report of Agro-forestry Department of Acharya NG Ranga Agricultural University 2012, Hyderabad, ICAR-Central Research Institute for Dryland Agriculture and Andhra Pradesh State Forest Department, Hyderabad.

Anonymous 2012A. https://www.easyayurveda.com/2012/12/21/karanja-pongamia-pinnata-benefits-usage-ayurveda-details/

Arathi HS, Ganeshaiah KN, Umashaankar R. and Hegde SG 1999. Seed abortion in Pongamia pinnata (Fabaceae) *American J. Botany*, **86:** 659-662.

Arpiwi NL, Guijun Yan, Elizabeth L, Barbour J, Plummer A 2014. Phenology, pollination and seed production of *Millettia pinnata* in Kununurra, Northern Western A ustralia, *Journal Biologi*.**18:** 19-23.

ARS-USDA 2005. Developing and /or use molecular tools to identify and clone desirable genes from regenerated plants. https://portal.nifa.usda.gov/web/crisprojectpages/0409866-develop-andor-use-molecular-tools-to-identify-and-clone-desirable-genes-from-rangeland-plants.html.

Arvind and Bekal S 2018. Experimental Investigation on the Performance of Novel Stove for Use with Vegetable Oil, *Energy and Power* **8:** 46-50.

Athaya CD1985. Ecological studies of some forest tree seeds. II. Seed storage and viability. *Indian J. Forestry,* **8:** 137–140.

Berk Z, 2013. Food Process Engineering and Technology (Second Edition). *Advances in Molecular Toxicology,* **2014**. ISBN 978-0-12-415923-5.

Bhatt RP, and Sanjappa M 1976. Cytology of Pongam oil tree *Derris indica* Lamk. Bennett. *Curr. Sci.,* **45:** 388–389.

Biswas B, Kazakoff SH, Jiang Q, Samuel S, Gresshoff PM, and Scott PT 2013. Genetic and genomic analysis of the tree legume *Pongamia pinnata* as a feedstock for biofuel. *Plant Genome* **6:** 1-15.

Bodade SN and Khyade VB 2012. Detailed Study on the Properties of *Pongamia pinnata (*Karanj*)* for Production of Biofuel. *Res.J. Chemical.Res.* **2:** 16-20.

CAB International 2000. Forestry Compendium Global Module, Wallingford, UK: *CAB International.*

Cervera M, Juarez J, Navarro A, Pina JA, Duran-Vila N, Navarro L, et al 1998. Genetic transformation and regeneration of mature tissues of woody fruit plants bypassing the juvenile stage. *Transgenic Res.* **7** 51–59.

CFTRI, 2012. Consolidated Report from Karanja (*Pongamia)* Seeds for VAYUGRID Market Place Services Pvt. Ltd. Bangalore *Central Food Technological Research Institute, CSIR,* Mysore (India) 18 pages.

Chaturvedi RK, Raghubanshi AS, and Singh JS. 2011. Carbon density and accumulation in woody species of tropical dry forest in India *Forest Ecology and Management,* **262:** 1576–1588

Chipade VV, Tankar AN, Pande VV., Tekade AR, Gowekar NM, Bhandari SR, and Khandake SN 2008. Phytochemical constituents, traditional and pharmacological properties. A Review. *Int. Green Pharm* **2:** 72-75 http://www.greenpharmacy.info/text.asp?2008/2/2/72.

Chodhury RR, Basak, S, Ramesh AM, and Rangan L 2013. Nuclear DNA content of *Pongamia pinnata* L. and genome size stability of *in vitro* regenerated plantlets. *Protoplasma* **251:** 703–709.

CRIDA 2012 Progress Report 2012.*Central Research Institute of Dryland Agriculture*, ICAR. Hyderabad India.

Daniel J N 1997. *Pongamia pinnata*: a nitrogen fixing tree for oilseed. FACT 97-03, Forest Farm and Community Tree Network, USA.

Daisy Thomas, Naris Pal, Chawda SS, and Subba Rao K 2001. Bee-flora and Migratory Routes in India. *In Proc. 37th Int. Apic. Congr.*Durban, South Africa.

Devaraj R, and Sundararaj R 2014. Parasitoids of *Asphondylia pongamiae* (Diptera: Cecidomyiidae), the flower gall inducer of *Pongamia pinnata* and their roles in biological control. *Journal of Tropical Forest Science* **26:** 173-177.

Divakara BN, and Das R 2011.Variability and divergence in *Pongamia pinnata* for further use in tree improvement. *J. Forest Res.* **22:** 193-200.

Dsilva E 2015. *Down To Earth.* June, 07. London, UK.

Duke JA 1981. *Handbook of legumes of world economic importance.* New York, Plenum Press USA.

Duke JA 1983. *Handbook of Energy Crops. Wikipedia.*

Dwivedi G, and Sharma MP 2014 Prospects of biodiesel from *Pongamia* in India. *Renew Sust. Energ. Rev.* **32** 114-122.

Elevitch CR, and Wilkinson KM 1999. Nitrogen fixing Tree Startup Guide. *Sustainable Agricultural Research. USDA. Permanent Agricultural Resources, Holualoa,* USA.

EPA 2000. *A Citizen's Guide to Phytoremediation.* EPA 542-F-98-011. United States Environmental Protection Agency Washington DC, USA.

Etim EE 2012. Phytoremediation and its mechanisms. A Review. *Int. J. Environ. Bioenergy.* **2** 120-136.

FAO 1999. State of World Forests. *Food and Agricultural Organization,* Rome, Italy.

Finkeldey R, Leinemann L, and Gailing, O. 2010. Molecular genetic tools to infer the origin of forest plants and wood, *Appl. Microbiol. Biotechnol.* **85** 1251–1258.

Gelvin SB 2003. Agrobacterium-mediated plant transformation: the biology behind the "gene-jockeying" tool. *Microbiol. Mol. Biol. Rev.* **67** 16–37.

Gera M, and Chauhan S 2010. Opportunities for Carbon sequestration benefits from growing trees of medicinal importance on farm lands of Haryana *Indian Forester,* **136** 287-300.

Gore VK, and Satyamoorthy P 2000. Determination of Pongamol and Karanjin in Karanja Oil By Reverse Phase Hplc, *Analytical Letters,* **33** 337-346.

Green My Life 2017. https://www.greenmylife.in/shop/planting, supplies/pongamia-cake.

Gresshoff PM 2014. The Contrasting Need for Food and Biofuel: can we afford biofuel? *A Love of Ideas Melbourne University Press*144-152 pages.

Gupta HS 2009. Forest as Carbon Sink: Temporal Analysis for Ranchi district. *Indian Forester,* **32** 7-11.

Hnatiuk RJ 1990. Census Australian Vascular Plants, *Bureau Flora and Fauna, Canberra* (Australia).

Jain SM, and Minocha S 2000. Molecular Biology of woody plants: **1**'Springer, Dordrecht

Jaiswal N, Yadav PP, Maurya R, Srivastava AK, Tamrakar AK 2011. Karanjin from Pongamia pinnata induces GLUT4 translocation in skeletal muscle cells in a phosphatidylinositol-3-kinase-independent manner. *Eur J Pharmacol.* **670:** 22-28

Jesse M, 2009. Carbon Sequestration Potential of the Million Trees. NYC Initiative, *Biofuels and Bio-Based Carbon Mitigation,* https://snrecmitigation.wordpress.com/?s=Carbon+Sequestration+Potential+of+the+ Million+Trees

Jiang Q, Yen SH, Stiller J, Edwards D, Scott PT, and Gresshoff PM 2012. Genetic, biochemical, and morphological diversity of the legume biofuel tree *Pongamia pinnata. J Plant Genome Science* **1:** 54-68.

Jin Y, Liu L, Hao X, Harry DE, ZhengY, Huang,T, and Huang J 2019. Unravelling the MicroRNA-Mediated Gene Regulation in Developing *Pongamia* Seeds by High-Throughput Small RNA Profiling. *Int. J. Mol. Sci,* **201:** 3509

Judy L, Souvannavong O, and Dawson IK 2014. Quintessentially seeing the trees as well as the forest: The importance of managing forest genetic resources *Forest Ecology and Management,* **333:** 1-8.

Kamarkar A, Kamarkar S, and Mukherjee S 2012. Biodiesel production from neem towards feed stock diversification, *Renewable and Sustainable Energy Reviews,* **16** :1050–1060.

Karen DH, Rakan A, Zahawi A, Cole RJ, Ostertag R, and Corde S 2010, Planting Seedlings in Tree Islands Versus Plantations as a Large-Scale Tropical Forest Restoration Strategy. Restoration Ecology, *Journal of Society for Ecological Restoration International, Society for Ecological Restoration International* **19** 470-479.

Karikalan L, and Chandrasekaran M 2015. Karanja oil biodiesel: A potential substitution for diesel fuel in diesel engine without alteration, *ARPN Journal of Engineering and Applied Sciences* **10:** 152-161.

Karmee SK, and Chadha A 2005 Preparation of biodiesel from crude oil of *Pongamia pinnata. Bioresour Technol.* **96:** 1425-1429.

Karoshi VR, and Hegde GV 2002 Vegetative Propagation of *Pongamia pinnata* L Pierre: Hitherto a Neglected Species *Indian Forester* **128:** 3.

Kazakoff SH, Imelfort, M., Edwards, D., Koehorst, J., Biswas, B., Batley, J., Scott, P.T. and Gresshoff, P.M. 2012. Capturing the biofuel wellhead and powerhouse: the chloroplast and mitochondrial genomes of the leguminous feedstock tree Pongamia pinnata. PloS One December 2012 | Volume 7 | Issue 12 | e5168.

Kelly AJ, Zagotta MT, White RA, Chang C, and Wagner MDR 1990. Identification of genes expressed in the tobacco shoot apex during the floral transition. *The Plant Cell* **2** 963-972.

Kesari L, Anitha K, and Rangan L 2008. Systematic characterization and oil analysis in candidate plus trees of biodiesel plant *P. pinnata*. *Annals of Applied Biology* **152:** 397-404.

Kesari V, Krishnamachari A, and Rangan L 2008. Systematic characterisation and seed oil analysis in candidate plus trees of biodiesel plant, *Pongamia pinnata Annals of Applied Biology*. **152:** 397–404.

Kesari V, Ramesh A M, and Rangan L 2013. Characterization of root nodulating bacteria from biodiesel crop, *Pongamia pinnata* L. *Bio Med. Research International* dx.doi.org/10.1155/2013/165198

Kesari V, Ramesh VA, and Rangan L 2013. A *Rhizobium pongamiae* sp. nov. from Root Nodules of *Pongamia pinnata., BioMed Research International* / 2013 165-198

KesariV, and Rangan L 2010. Development of *Pongamia pinnata* as an alternative biofuel crop- current status of plantations in India and scope. *Journal of Crop Science and Biotechnology* **13:**127-137.

Kesari V, Ramesh AM, and Rangan L 2012. Rapid multiplication and assessment of genetic purity by biomolecular techniques in micropropagated plants of *Pongamia pinnata*, a potential biodiesel plant *Biomass and Bioenergy* **44:** 23-32.

KesariV, and Rangan L 2012 Electrophoretic patterns of proteins isolated from immatured and matured stages of 10 candidate plus trees of versatile oleaginous legume, *Pongamia pinnata* L Pierre. *Agroforestry Systems* **84:** 157-161.

Kesari V, Vinod MS, Parida A, and Rangan L 2010. Molecular marker based characterization in candidate plus trees of *P. pinnata*, a potential biodiesel legume from North Guwahati, Assam. *Aob Plants* **2010** 17 DOI 10.1093/aobpla/plq017

Korwar G R, Prasad JVNS, Rao G R, Venkatesh G, Prathibha G, and Venkateswarlu B. 2014. Agroforestry as a strategy for Livelihood Security in Rainfed Areas.: Experience and Expectations. In *Agroforestry Systems in India: Livelihood Security and Ecosystems Services Advances in Agroforesttry***10** [eds] J.C. Dagar et al *Springer* India.

Kumar S, and Chadha K A 2005 Preparation of biodiesel from crude oil of *Pongamia pinnata. Bioresour. Technol.* **96:** 1425-1429 https:// doi.org/10.1016/j.biortech. 2004. 12.011.

Kumar S, Mehta UJ, and Hazra S 2009. *In vitro* studies on chromium and copper accumulation potential of *Pongamia pinnata* L. Pierre seedlings. *Bioremediation Biodiverse Bioavailab.* **3:** 43-48.

Kumar D, Singh, B and Sharma YC 2017. Bioenergy and Phytoremediation Potential of *Millettia pinnata,* Chapter 6 In *Phytoremediation of Bioenergy plants* (eds) K Baudh , B Singh, and Korstad *Springer* 169-188.

Kumar S, Radhamani J, and Srinivasan K 2011. Physiological and biochemical changes in seeds of karanj (Pongamia pinnata) under different storage conditions Indian *J. Agric. Sci.*, 81 5.

Kumar, Vishal K, Chandrashekar, and Om Prakash Sidhu 2006 Efficacy of karanjin and different extracts of *Pongamia pinnata* against selected insect pests. *J. Ent. Res.,* **30** (2): 103-108 (2006)

Lale A, and Kulkarni DK 2010. A mosquito repellent karanja kunapa from *Pongamia pinnata. Asian Agri.History.* **14:** 207-211.

Low, T and Booth, C 2008. The weedy truth about biofuels, Invasive Species Council, Melbourne.

Ludwig D, Mangel M, and Haddad B 1997. Ecology, Conservation and Public Policy. *Ann. Rev. Ecol. Syst.* **32:** 481–517.

Lugo AE 1997. The apparent paradox of re-establishing species richness on degraded lands with tree monocultures. *Forest Ecology and Management,* **99**: 9-19.

Majumdar D, Pandya B, Arora A, and Dhara S 2004. Potential use of karanjin (3-methoxy furano-2,3,7,8-flavone) as nitrification inhibitor in different soil types. *Archives of Agronomy and Soil science.* **50:** 455-465.

Manoharan S, and Kour J 2013. Anti-cancer, anti-viral, anti-diabetic, anti-fungal and phyto-chemical constituents of medicinal plants. *Am. J. Pharm. Tech. Res*. **3** 150-169.

Mao J, He Z, Hao J, Liu T, Chen J, and Huang S 2019. Identification, expression, and phylogenetic analyses of terpenoid biosynthesis-related genes in secondary xylem of loblolly pine *Pinus taeda* L based on transcriptome analyses *Plant Biology*, **2013** *PeerJ* **7**:e6124PubMed 30723613 https://doi.org/10.7717/peerj.6124

Mittler R, and Blumwald E 2010. Genetic engineering for modern agriculture: challenges and perspectives *Annual. Rev. Plant Biol.* **61:** 443–462.

Morton JF 1990. The pongam tree, unfit for Florida landscaping, has multiple practical uses in under-developed lands', *Proceedings of the Florida State Horticultural Society* **103:** 338–43.

Mukta N, Murthy, IYLN, and Sripal P 2009. Variability assessment in *Pongamia pinnata* L Pierre, germplasm for biodiesel traits. *Industrial Crops and Products* **29:** 536–540.

Mukta N, and Sreevalli Y 2010. Propagation techniques, evaluation and improvement of the biodiesel plant, *Pongamia pinnata* L.Pierre—a review *Industrial Crops and Products* **31:** 1–12

Nasareen PNM, Varadan SY, and Ramani N 2013. Damage assessment of gall mite *Aceria pongamiae* Keifer 1966 *(Acari: Eriophyide)* on *Pongamia pinnata* (L). Pierre. In *Prospects in Biosciences: Addressing Issues.* **2013** 325-333 A Sabu and A Augastine [eds] Springer India.

Negi KS, and Tiwari CK 1984. Vegetative propagation in cuttings of *Pongamia pinnata, Indian Forester* **110:** 655-659.

Neo P 2020. Going Carbon negative: TerViva establishes world- first organic Pongamia bean supply chain in India. *Food navigator-asia.com.,* Aug.04, 2020. https://www.foodnavigator-asia.com/Article/2020/08/04/Going-carbon-negative-TerViva-establishes-world-first-organic-pongamia-bean-supply-chain-in-India.

Orwa C, Mutua A, Kindt R, Jamnadass R, and Simons, A 2009. Agro-forestry Database: a tree reference and selection guide. Version **4.0** 1-6. http://www. worldagroforestry.org/treedb2/ AFTPDFS/Pongamia_pinnata.pdf

Osman M, Wani SP, Balloli T, Sreedevi K, Srinivasa Rao CH, and D'Silva E 2005. *Pongamia* Seed Cake as a valuable source of plant nutrients for sustainable agriculture. *Indian. J. Fert.* **5** 25-26&29-32.

Palanisamy K, Ansari SA, Pramod Kumar, and Gupta BN 1998. Adventitious rooting in shoot cuttings of *Azadirachta indica* and *Pongamia pinnata. New Forests* **16:** 81–88.

Panda AK, Sastry VRB, Kumar A, and Saha SK 2006. Quantification of karanjin. Tannin and trypsin inhibitors in raw and detoxified expeller and solvent extracted karanj (*Pongamia glabra*) cake. *Asian Aust. J Anim. Sci.* **19:** 1776-1783.

Pandey A 2008. *Hand Book of Plant-based Biofuels,* CRC 255-266. ISBN156022 175 5.

Pandey VC 2017. Managing waste dumpsites through energy plantations *In* KS Bauddh S Bhaskar, and J Korstad [eds] *Phytoremediation Potential of Bioenergy Plants*, 1st ed., Springer, 2017, p. 15.

Parisi C, Tillie P, and Cerezo RE 2016. The global pipeline of GM crops out to 2020 *Nat. Biotechnol* **34** 32–36.

Patel J S and Narayana GV 1937. Chromosome numbers in some economic flowering plants., *Current Science* 1937.

Pavela R,and Herda G 2007. Repellent effects of *Pongamia* oil on settlement and oviposition of the common greenhouse whitefly *Trialeurodes vapurariorum* on Crysanthemum *Insect Science* **14** 219-224.

Pavithra HR, Shivanna MB, Kumar, KR, Prasanna KT, and Gowda B 2015. Morphogenetic diversity of *Pongamia pinnata* L Pierre candidate plus trees. A potential agroforestry tree *Int J Sci Nat.* **6:** 341-352.

PérezJiménez M, Carrillo-Navarro A, and CosTerrer J 2012. Regeneration of peach *Prunus persica* L. Batsch cultivars and *Prunus persica× Prunus dulcis* rootstocks via organogenesis *Plant Cell Tissue Organ Culture* **108:** 55–62.

Phartyal SS, Thapliyal RC, Koedam N, Godfroid S 2002. *Ex situ* conservation of rare and valuable forest tree species through seed-gene bank. *Curr. Sci.*, **83**:1351–1357.

Prasad MNV 2007. Phytremediation in India *In* N Willey [ed] *Methods and Reviews.* Humana Press, Totowa, NJ USA 435-454.

Prasad MVR 1994. Minor Oil bearing Species of Forest Origin for Diversification of Vegetable Oil Production. *In* MVR Prasad, et al [eds] *Sustainability in Oilseeds. Indian Society of Oilseeds Research, Directorate of Oilseeds Research, Hyderabad.* 91-98.

Prasad MVR 2012a. *Pongamia* Based Sustainable Bio-energy System *J. Ecosyst. Ecogr.* **2:** 65.

Prasad MVR 2012b. Sustainable Bio-energy through *Pongamia* VayuSap. *2nd National Conference on Recent Advnces in Bio-energy. Ministry of New & Renewable Energy Dec. 7-8, 2012. Kapurthala Punjab India* **43**.

Prasad MVR 2013. Genetic enhancement of *Pongamia* for Bio-energy, *Recent Advances in Bioenergy Research- Vol.III SSS National Institute of Renewable Energy, Govt. of India. Kapurthala, Punjab http://www.nire.res.in/e-Book%20III.pdf*

Prasad MVR 2019. Environmental Amelioration through *Pongamia pinnata* based Phytoremediation, *International Journal of Science and Research* **8:** 93-102.

Prasad MVR, Langa A, and Consolo JP 2000. Selection of superior genotypes of cashew in Nampula zone of Mozambique. *The Cashew* Jan-March,200, 8-23.

Prasad MVR, and Vijaykumar Y 2016. Energy Potential of Genetically Elite Clones of *Pongamia pinnata, International Conference on Recent Advances in Bio-energy Research, Ministry of New & Renewable Energy, Sardar Swaran Singh National Institute of Renewable Energy, Govt. of India,* Kapurthala India, Feb.25-26, 2016.

Punitha R *et al* 2006. Effect of *Pongamia pinnata* flowers on blood glucose and oxidative stree in alloxan induced diabetic rats. *Indian J Pharmaco* **38:** 62-63.

Qunyi, Jiang, Shang-Heng, Yen, Jiri Stiller, David Edwards, Scott PT, and Gresshoff PM 2012. Genetic and Morphological Diversity of the Legume Biofuel Tree *Pongamia pinnata, Journal of Plant Genome Sciences* **1:** 54–67, 2012.

Rabab R, and Dixit Madhurima 1997/ *J. American Oil Chemists Society*, **74P:** 9.

Raghavan RS, and Arora CM 1958. Chromosome numbers in Indian medicinal plants-II, *Proceedings of the Indian Academy of Sciences - Section B.*, **47** 352–358

Rai M K, and Shekhawat, NS 2014. Recent advances in genetic engineering for improvement of fruit crops. *Plant Cell Tissue Organ Cult.* **116:** 1–15.

Raju AJS, and Rao SP 2006. Explosive pollen release and pollination as a function of nectar feeding activity of certain bees in bio-diesel plant *Pongamia pinnata* Pierre (*Fabaceae). Curr.Sci.* **90:** 960-967.

Ramesh AM, Bask S, Roychoudhury R,and Rangan, L 2014. Development of flow cytometric protocol for nuclear DNA content and determination of chromosome number in *Pongamia pinnata* L. *Appl.Biochem. Biotechnol.* **172** 533-548..

Ramesh AM, Basak S, Choudhury RR, and Rangan L 2013. Flow cytometric estimation of nuclear DNA content and determination of mitotic chromosome count in *Pongamia pinnata*, a valuable biodiesel plant. *Applied Biochemistry and Biotechnology* DOI: 10.1007/s12010-013-0553-z.

Ranjan, S 1977. Morphogenic studies on *Derris indica* (Lamk) Bennet. *Indian Forester* **123:** 837–839.

Reddy NS, Ramesh G, and Suryanarayana, B 2009 Evaluation of Tree species under different land use systems for higher Carbon sequestration. *Indian J. Dryland Agri. Res. & Dev.* **24:** 74-78.

Reiner Finkeldey R, Leinemann L, and Oliver Gailing O 2010. Molecular genetic tools to infer the origin of forest plants and wood, *Appl. Microbiol. Biotechnol.* **85:** 1251–1258.

Rout Sandeep and Nayak Saswat 2015. Effect of Storage Conditions and Duration on Seed Germination of *Pongamia pinnata* L.Pierre *Research Journal of Agriculture and Forestry Sciences* ISSN 2320-6063., **3**(4):19-24, April (2015) Res. J. Agriculture and Forestry Sci. International Science Congress Association 19.

Rout S, and Nayak, S 2015. Vegetative propagation of karanja (*Pongamia)* through stem cuttings *Journal of Applied and Natural Science* **7:** 844-850.

Sabbadini, S., Pandolfini, T., Girolomini, L., Molesini, B., and Navacchi, O. 2015. "Peach (*Prunus persica* L.)" *Agrobacterium Protocols*, Wang K (ed) New York : Springer 205–215.

Sahoo DP, Rout GR, Das S, Aparajita S, and Mahapatra AK 2011. Genotypic variability and Correlation Studies in Pod and Seed Characteristics of *Pogamia pinnata* L Pierre in Orissa: India *Int J Forestry Re* **2011** 1-6 ID 728985.

Samuel S, Scott PT, Gresshoff PM 2013. Nodulation in the legume biofuel feedstock tree *Pongamia pinnata. Agricultural Research* DOI 10.1007/s40003-013-0074-6.

Sangwan S, Rao D, and Sharma RA 2010. A Review on *Pongamia Pinnata* L Pierre A Great Versatile Leguminous Plant *Nature and Science* **8:** 11.

Sarbhoy RK 1977. Cytogenetical Studies in *Pongamia pinnata* L Pierre *Cytologia* **42:** 415-423.

Sedjo RA, and Botkin D 1997. Using forest plantations to spare natural forests. *Environment.* **30** 15-20.

Selvi B, and Kadamban D 2009. Studies on the parasitic plants of Pondicherry Engineering College Campus, Puducherry *International Journal of Plant Sciences* **4:** 547-550.

Sengupta RP 2013. Ecological Limits and Economic Development. *Oxford University Press,* New Delhi.

Shelke G 2010. Genetic Characterization of *Pongamia Pinnata* Populations Muller EK (ed) *VDM Verlag* ISBN10 & 13, 56 pages.

Shirbate N and Malode SN 2012. Heavy metals phytoremediation by *Pongamia pinnata* L growing in contaminated soil from municipal solid waste landfills and compost Sukali Dpot Amravati Maharashtra *International Journal of Advance Biological Research* **2:** 147–152.

Shrinivasa U 2001. Honge *Pongamia* Oil proves to be a good biodiesel. *http://goodnewsindia.com/Pages/content/discovery/honge.html*

Singhal VK, Gill BS, and Sidhu MS 1990. Cytological explorations of Indian woody legumes *Proceedings Plant Sciences* **100:** 319-331

Soren, Sastry VRB, Saha SK, Wankhade UD, Lade MH, and Kumar A 2009. Performance of growing lambs fed with processed kranja *Pongamia glabra* oilseed cake as partial protein supplement to soybean meal *J Animal Physiol Animal Nutri* **93:** 237-244.

Srinivas I, RajeswarRao G, Pratibha G, Korwar GR, and Venkateswarlu B 2009. Effect of operating conditions on oil extraction from *Pongamia pinnata* seeds for on-farm value addition *J. Oilseeds Res* **26:** 700–702.

Sugla T, Purukayastha J, Singh SK, Solleti SK, Sahoo L 2007. Micropropagation of *Pongamia pinnata* through enhanced axillary branching. *Invitro Cell Dev.Biol. Plant* **43:** 409–414.

Sujatha K, Hazra S 2006. *In vitro* regeneration of *Pongamia pinnata* Pierre. *J Plant Biotechnol* **33:** 263–270.

Sujatha K, Hazra S 2007. Micropropagation of mature *Pongamia pinnata* Pierre *Invitro Cell Dev Biol Plant* **43:** 608–613.

Sujatha K, Panda BM, Hazra S 2008. De novo organogenesis and plant regeneration in *Pongamia pinnata*, Oil producing tree legume. *Trees* **22:** 711–716.

Sundararaj R Rajamuthukrishnan, and Remadevi OK 2005. Annotated list of insect pests of *Pongamia pinnata* L Pierre in India *Annals of Forestry* **13:** 337–341.

Sunil N, KumarV, Sivaraj N, Lavanya C, Prasad RBN, Rao BSK, and Varaprasad KS 2010. Variability and divergence in *Pongamia pinnata* (L) Pierre germplasm. - a candidate tree for biodiesel. *Global Change Bioenergy,* **1:** 382-391.

Sweet B, and Mezzetti B 2017. New Biotechnological Tools for the Genetic Improvement of Major Woody Fruit Species *Front Plant Sci* **15:** https://doi.org/10.3389/fpls 2017,01418

Tamarkar AK, Yadav PP, Tiwari P, and Maurya R 2008. Identification of pongamol and karanjin as lead compounds with anti hyperglycemic activity from *Pongamia pinnata* fruits. 2 *Journal of Ethnopharmacology* **118:** 435-439.

Terviva 2006. Frequently Asked Questions FAQ About *Pongamia. Terviva,* USA. *https://www.crunchbase.com/organization/terviva*

TerViva 2020. *Pongamia* trees to revitalize fallow Citrus acreage in Florida, *Feed Stuffs. This Week in Agribusiness, Sept. 26* https://www.feedstuffs.com/nutrition-health/pongamia-trees-examined-protein-meal-potential

Thomas D, Pal N, Chawda S S and Rao K S 2001. Bee flora and migratory routes in India, *Proc 37th Int Apic Congr,* Oct.28- Nov.1, 2001, Durban, South Africa.

Tulod AM, Castillo ASA, Carandang WM, and Pampolina NM 2012. Growth performance and phytoremediation potential of *Pongamia pinnata* (L) Pierre, *Samana saman &Vitex parviflora* copper contaminated soils, amended with Zeolite and VAM. *Asia Life Sciences* **21:** 499-522.

USAID 2007. Vegetative propagation techniques, Perennial Crop Support Series, Jalalabad, Afghanistan. *Publication No.2007-003-AFG.* Nov.18, 2007, 38 pages https://www. sas.upenn.edu/~dailey/VegetativePropagationTechniques.pdf

USDA NRCS 1999. The PLANTS database, National Plant Data Centre, Baton Rouge, LA 70874- 4490 United States.

Varaprasad KS 2014. Oilseeds Scenario in India and Way Forward. *Souvenir, Indian Oilseeds Produce Export Promotion Council,* Mumbai.

Varaprasad K S, and Sudhakara Babubu SN 2015. Technologies for Increasing Oilseeds Production, *Lead paper on Technologies for Enhancing Oilseeds Production through NMOOP, National Seminar, Hyderabad, January* 18-19.

Vayugrid 2014. Progress Report. Board of Directors, *Vayugrid Market Place Services Pvt. Ltd.* Bangalore, India. 21pages.

Vinay BJ, and Sindhukanya TCS 2008. Effect of detoxification on the functional and nutritional quality of proteins of karanja seed meal, *Food Chem.* **106:** 77-84

Vismaya W, Sapna EJR, Manjunatha P, Srinivas TC, and Sindhukanya 2010. Extraction of karanjin: A value addition to karanja *(Pongamia pinnata*) seed oil, *Industrial Crops and Products* **32:** 118-122.

Vismaya W, Srikanta M, Belagihally, Sindhu Rajashekhar, Jayaram VB, Shylaja M, Dharmesh, and Sindhukanya TCS 2011. Gastro-protective Properties of Karanjin from Karanja (*Pongamia Pinnata*) Seeds, Role as Antioxidant and H+,K+-ATPase Inhibitor) *Evidence-Based Complementary and Alternative Medicine* **2011**, ID 747246, 10 pages.

Vivek, and Gupta AK 2004. Biodiesel production from Karanj oil, *Journal of Scientific and Industrial Research,* **63:** 39-47.

Wani SP, and Sreedevi TK 2005. Pongamia's journey from forest to micro-enterprise for improving livelihoods. *International Crops Research Institute for the Semiarid Tropics, Patancheru,* Andhra Pradesh, India, p. 4.

Wani SP, Sreedevi TK 2007. Strategy for rehabilitation of degraded lands and improved livelihoods through biodiesel plantations. *Proc. 4th International Biofuels Conf. Winrock International, New Delhi,* 50–64.

Warr Benjamin 2018 Mapping the opportunity. Green Mine Zones, Copper Belt Zambia. *Better World Energy Ltd.* Zambia. 60 pages.

WRI 1998 *World resources Institute.* 1998-99 https://www.wri.org/A guide to the global environment.

Yadav RD, Jain SK, Alok S, Kailasiya D, Kanaujia VK, and Kaur S 2011. A Study on Phytochemical Investigations of *Pongamia pinnata*, Linn. Leaves *IJPSR* **2:** 2073-2079.

Yadav RD, Jain SK, Alok S, Prajapati SK, and Verma A 2019. *Pongmia pinnata*: An Overview, *International Journal of Pharmaceutical Sciences and Research* A Web of Science - ESCI Indexed Journal Projected Impact Factor 2019 1.230.

Zhuang P, Ye ZH, Lan CY, Xie ZW, and Hsu WS 200 Chemically assisted phytoextraction of heavy metal contaminated soils using three plant species. *Plant Soil* **276:** 153-162.